VITAL HARMONIES
Molecular Biology and Our Shared Humanity

Erwin Fleissner

Columbia University Press   New York

Columbia University Press
Publishers Since 1893
New York Chichester, West Sussex
Copyright © 2004 Columbia University Press

Library of Congress Cataloging-in-Publication Data
Fleissner, Erwin.
Vital harmonies : molecular biology and
our shared humanity / Erwin Fleissner.
p. cm.
Includes bibliographical references and index.
ISBN 0–231–13112–7 (cloth : alk. paper)
1. Molecular biology—Popular works.
2. Molecular biology—History—Popular works.
I. Title
QH506.F545   2004
572.8—dc22   2004041313

Columbia University Press books are printed
on permanent and durable acid-free paper.

Printed in the United States of America
Designed by Audrey Smith
c 10 9 8 7 6 5 4 3 2 1

The nature of everything is dual—matter
And void; or particles and space, wherein
The former rest or move. . . .
And death does not destroy the elements
Of matter, only breaks their combinations.

—Titus Lucretius Carus (ca. 99—ca. 55 B.C.),
On the Nature of Things, trans. Rolfe Humphries

Sufferings are not recognized,
Love is not learned,
And what in death makes us distant
Is not unveiled.

—Rainer Maria Rilke (1875–1926), *Sonnets to Orpheus*

CONTENTS

PREFATORY NOTE

The reference to Lucretius on page v of this book is not incidental. Among those who followed in the footsteps of Democritus (460—370 B.C.) and Epicurus (341—270 B.C.), Lucretius had a unique gift, as a poet, of conveying to his readers how seeing life as arising from matter itself ("atoms in the void") need not diminish the richness and variety, the depth and *human quality* of existence. Thus he encompassed in his masterpiece, *De Rerum Natura* (*On the Nature of Things*), not only cosmic processes and natural wonders but also the substance and texture of everyday life—friendship, love, birth, and, yes, death. His narrative ends unexpectedly, not with some evocation of earthly bliss but with stanzas picturing the great plague of Athens (430—429 B.C.). It is as though he were saying to us, across two millennia, "this too has its explanation—not in the realms of superstition or divine retribution—but as hard fact that can be grasped by the human mind."

The lines from Rilke on page v (from *Sonnets to Orpheus,* part 1, 19; my translation), written not too long before his death from leukemia in 1926, are there to convey something of the irreducible mystery of sheer existence—of *being* alive.

Finally, let me add that I have no intention (if it were possible!) of presenting and explaining the whole of molecular biology—as if this were some sort of highly condensed yet all-encompassing textbook. For the interested there do exist excellent, well-crafted textbooks, particularly on defined subareas of this vast subject, written at various levels of sophistication; there are also books that capture important aspects of the history of the field (see the select bibliography and further reading). What is described in the pages that follow is something in the nature of a personal odyssey. I hope that the prologue and the final chapter of this book convey some of the reasons that motivated me to write it after the century, just ended, of scientific revolutions, as Thomas Kuhn called them, which has left us, as a species, with a sense of human solidarity, concern for our planet, anticipation, hope, foreboding, uncertainty in the face of what is unknown, and awe at the existence of life.

ACKNOWLEDGMENTS

For conversations and written communications concerning scientific matters covered in this book, I am indebted to James Darnell, Michael Young, Ali H. Brivanlou, and Kathryn Zimmerman at the Rockefeller University. Also Harold Varmus, president, and Lee Niswander, Joan Massagué, and Jerome Posner at Memorial Sloan-Kettering Cancer Center. At Hunter College of the City University of New York: in Biological Sciences, Peter Lipke and Jesus Angulo; in Chemistry, Richard Franck and Pamela Mills; in Physics, Steven Greenbaum; and in Computer Sciences, Susan Epstein. At New York University, in Biology, Claude Desplan. At the American Museum of Natural History, in the Division of Anthropology (with his focus on primate and human evolution), Ian Tattersall. At the Scripps Research Institute, La Jolla, California, Peter Vogt.

For creating line drawing illustrations for the book the cooperation and skill of Dimitri Drjuchin was much appreciated. Teresa Kruger of the Cold Spring Harbor Laboratory Archives was very helpful in identifying photographs for use as figures. The photograph of Fritz Lipmann is from Robin McElheny of the Harvard University Archives. For researching data relevant to the scientific drawings in the figures (e.g., scale bars), especially by use of the Internet, my thanks go to Kathryn Ray.

John Heuston and his expert word-processing staff at Riverside Copy, 248 West 106th Street, New York, provided reliable assistance as I underwent a prolonged, clinically diagnosed, vision impairment.

It was a pleasure to work with my editor at the Columbia University Press, Robin Smith. His philosophical attitude, seasoned with practical advice and a ready sense of humor, has been what I needed to make this book, so long in gestation, a reality. I also benefited from the expertise, ingenious suggestions, and good humor of my manuscript editor at the Press, Susan Pensak.

In a very personal sense I was influenced in the original conception of this book by an invitation, over two decades ago, to address students and faculty at Lander College in South Carolina. The invitation was made by the college's then president, Larry Jackson, and was arranged by my lifelong closest friend, Robert Cumming, and his wife, Deborah, both faculty members at Lander. My title for the talk was "The Credo of a Biologist," and in the preparation of the lecture I was moved to consider some of the very ideas that are developed in this book. My friend of many decades, André Schiffrin, founder of the New Press in New York, also offered some professional advice, as I began to search for a publisher. And, when final acceptance for publication still hung in the balance, my friend since Oxford days, the noted historian of physics Russell McCormmach, provided valuable encouragement, based on his reading of the manuscript.

Finally, my gratitude goes to my wife, Judith Friedlander, for her overall support, encouragement, and advice, from personal experience, on the vicissitudes of scholarly authorship, and her persistent optimism about final outcomes. She and I met after the tragic death of my first wife, the author Norma Klein, and I feel blessed to have found her. The personal support of my two—now adult—daughters, Jennifer and Katherine, has also meant a great deal to me.

Line art credits:

Figure 6 is redrawn from figure 1 of E1 of *Instant Notes: Molecular Biology*, 2d ed., by P. C. Turner, A. G. McLennan, A. D. Bates, and M. R. H. White (Oxford: Bios Scientific, 2000). Figures 8, 9, and 16 are redrawn from figures 1.4, 14.10, and 2.16, A, C, B of *Cell Biology* by T. D. Pollard and W. C. Earnshaw (Philadelphia: Saunders, 2002). Figure 10 is redrawn from figure 3.15 of *Molecular Biology of the Cell* by B. Alberts, D. Bray, J. Lewis, M. Raff, K. Roberts, and J. D. Watson (New York and London: Garland, 1983). Figure 14 is redrawn from figure 13.5 of *The Cell: A Molecular Approach* by G. M. Cooper (Washington, D.C. and Sunderland, Mass.: ASM and Sinauer, 1997).

Figure 11 is the author's design, based to some extent on figure 18.5 of Pollard and Earnshaw's *Cell Biology*. Figure 15 is the author's design.

Figure 13 is redrawn from an illustration by Sue Lauter in *The First Hundred Years, Cold Spring Harbor Laboratory* (Cold Spring Harbor, N.Y.: Cold Spring Harbor Laboratory, 1988).

VITAL HARMONIES

PROLOGUE

MOLECULAR BIOLOGY FROM THE
VIEWPOINT OF A PRACTITIONER

It was in the summer of 1987. Lying in the hospital the night before a brain operation—the second one in ten days, this time on an emergency basis, to address effects of hemorrhaging after the first— I looked back on a life of over fifty years: a career in scientific research (molecular biology, origins of cancer), marriage to a novelist, two daughters nearing adulthood, friends, including other scientists and writers, opportunities to visit scenes of natural beauty, physical pleasures, aesthetic fulfillment (music!), emotional pain and stress, exertion of work and the depth, clarity, and austere beauty of scientific understanding. "What a walk through life!" I thought, fascinated by this vision of abundance, not willed, or expected. The brain tumor had been the size of a fist, benign, and fortunately operable—the first operation had lasted over twelve hours; the hemorrhaging was traceable to an adverse drug reaction that destroyed my blood platelets. After more than a quarter century of laboratory experimentation, over twenty of those years in cancer research, I was now across the street in a hospital bed at the same cancer center (Memorial Sloan-Kettering in New York), an experiment of nature.

Our rational understanding of nature and our place in it is the— as yet unfinished—achievement of the scientific revolution of the last few centuries. The physical sciences dominated in the early stages, with some notable exceptions (for example Vesalius and Harvey in anatomy and physiology), and it is only in the last century and a half that biology has made broad and rigorous connections with the universe of natural laws. And it is the last half of the century just passed that has witnessed the dramatic union of biology with chemistry and physics in the new discipline of molecular biology. This union is the subject of this book—how it has happened, the hope that it has inspired, and the feeling of disquiet as well—for which the biblical apple may not be too far-fetched an analogy (or the hubris of Greek tragedy?).

Is there such a thing as too much knowledge about our own "inner workings"? Would we want to, if we could, put the genie back in the bottle and return to a simpler, more naive way of seeing things? The market for "organic" foods and remedies is suggestive. So is the hostility toward genetically altered foods (especially in Europe). But these are steps taken or not taken, with regard to food or remedial treatments. The more encompassing question is how we conceptualize ourselves. How do we fit into nature? Is nature itself something static? Can we suppress our—naturally evolved—mental capacities, and the body of knowledge to which it has led/is leading us, without dishonoring the very evolutionary process we seek to respect?

In the past forty-five years I have been able to observe and participate in the growth of this body of knowledge, and I am convinced that it is a natural development of the idealistic and humanistic traditions that are most praiseworthy in our civilization. If any civilization is defined by tradition (its continuity and identity) and information (ways of understanding, knowing how to do things), our science has vital links with both of these. Through these links it has supported life and creativity (Freud's *eros* in its broadest sense), without abolishing the possibility of destructiveness (his *thanatos,* or death instinct). Dostoyevsky's crabbed little man from "underground," sneering, "I say, gentlemen, hadn't we better kick over the whole show and scatter rationalism to the winds

. . . " still evokes something that is all too human. (Witness the bombings of public buildings and senseless attacks on young people in schools, in America and elsewhere, and human-against-human tragedies worldwide.)

More than a century ago Nietzsche proclaimed, outrageously, "God is dead!" This quintessential expression of skepticism rather than simple blasphemy was meant to force the discussion of truth, in its various forms, to start over again—at square one, so to speak. As Joseph Needham, the great British biologist (and also chronicler of science throughout Chinese history), pointed out many decades ago, there is no essential conflict of science with religion. There only appears to be when bureaucratic theologians invade the proper domain of science ("And still it moves . . . ," said Galileo, under his breath) or when scientists try to deduce moral principles from simple fact (eugenics never seems to give up the ghost).

The spirit of molecular biology, as a distinct branch of modern science, has much to do with the Cold Spring Harbor Laboratory on Long Island. Physically, especially in the early years, this institution was an informal-seeming collection of buildings, with some old houses, green lawns, and wooded areas adjacent to laboratories and conference facilities—all overlooking a small bay in Long Island Sound (figure 1). This was and is a center for research, pedagogy (its famed intensive "hands-on" summer courses on genetic experimentation at the frontiers of research), and exchange of ideas (half-week to weeklong conferences with hundreds of scientists, young and old, from around the world). The laboratory has seen the Nobel Prize awarded to two of its staff, Alfred Hershey and Barbara McClintock (for quite different advances), and, for the past thirty-five years, has had as its director, and now president, the Nobelist James Watson, famed for codiscovering the double-helical structure of DNA with Francis Crick, in collaboration with Maurice Wilkins—and Rosalind Franklin, who, as a member of Wilkins's lab, produced the actual X-ray data that supported the double helix model (figures 2 and 3).

FIGURE 1

Buildings of the Cold Spring Harbor Laboratory, as seen from the harbor on the north shore of Long Island. *Courtesy of the James D. Watson Collection, Cold Spring Harbor Laboratory Archives*

The hallmark of molecular biology (the term was coined just before 1960) is that, although its subject is the submicroscopic realm of molecules, its methods, seen in a historical perspective, have been highly visible, tangible, and straightforward—the sort of thing you could literally do on the kitchen table. (Granted, you might have had to order a few pieces of equipment.) The round petri dish, growing bacteria on the jellified surface of an agar-containing nutrient solution, was—and still is—the most time-honored item. More up-to-date gel devices of various shapes and sizes, the gel layers adhering to flat surfaces, are useful in separating and identifying molecules. Two-dimensional constructs or arrays of various materials can be helpful. Detection of radioactive or phosphorescent labeling by the use of film offers no great problem.

Let's look at some history. In the 1940s scientists at Cold Spring

FIGURE 2

James D. Watson, director of Cold Spring Harbor Laboratory, in 1975. *Courtesy of the James D. Watson Collection, Cold Spring Harbor Laboratory Archives*

FIGURE 3

Barbara McClintock of the Cold Spring Harbor Laboratory, in 1980, three years before she was awarded the Nobel Prize for her discovery of "mobile genetic elements." *Courtesy of the James D. Watson Collection, Cold Spring Harbor Laboratory Archives*

Harbor, especially Hershey and his occasional summer colleagues Max Delbrück and Salvador Luria, chose the viruses of bacteria, bacteriophages, to decipher the riddle of molecular genetics. Phage particles could reproduce hundreds of fold in host bacteria in less than an hour, lysing ("popping open") the bacteria. A single particle seeded on a "lawn" of bacteria in a petri dish would produce a round, clear spot in a few hours of infection and reinfection—one could count the seeded virus particles, at various dilutions, by counting such "plaques." Counterintuitively, most of the infecting phage particle seemed to remain stuck to the outside of the host bacterium. Yet, apparently something got in that carried information for the *bacterium* to make more phage—what was it?

Hershey and his co-worker Martha Chase (figure 4) did a classic experiment to explore this. They put radioactive phosphorus, ^{32}P (as phosphate), into a liquid culture of infected bacteria, collected the ^{32}P-labeled phage, allowed this to attach, again in liquid culture, to fresh bacteria and poured the culture into a household blender (then called a "Waring Blender" for historical reasons too complex to present here). They next turned on the blender to shear off the attached, stiff phage particles from the rather resilient bacteria and found that there was ^{32}P *inside* of the bacteria. It was known that ^{32}P labeled the DNA of the phage, so the implication was clear: phage DNA carried the information for making the next generation of phage. In the event Hershey and Chase had also labeled the phage with radioactive sulfur, ^{35}S, which got into the phage protein and got sheared off in the blender.

The "kitchen table" aspect of this experiment is obvious (the locale of virtually all blenders in those days of 1952). Add the availability of some radioactive isotopes (most probably by-products of the nuclear reactor at Oak Ridge), and a generous helping of ingenuity, and you get a result that was highly influential for the young James Watson and his British collaborators in choosing DNA as a prime suspect for being the material of genes. Received wisdom in those years had it that DNA, with its four chemical building blocks,[1]

FIGURE 4

Martha Chase and Alfred Hershey at the Cold Spring Harbor Laboratory in 1953, a year after the publication of the "blender experiment." *Courtesy of the James D. Watson Collection, Cold Spring Harbor Laboratory Archives*

was too simple to represent genes; proteins, with their choice of twenty constituents, being better candidates. The success of Oswald Avery and his colleagues in changing, "transforming," the character of bacteria by their exposure to purified, naked DNA (1944)— known to Watson and Crick—had not made as much of a splash as it should have. How pure was the DNA? muttered the doubters. Was there still some residual and crucial protein? The emphatic simplicity of the blender experiment (itself not 100 percent proof against doubting Thomases), intersecting the logic of the transformation experiment of Avery and his colleagues, made the case for DNA most attractive (figure 5).

Transformation by exposing cells to highly purified DNA has become routine in laboratories all over the world in the last half century. Conveniently, it isn't hard to get cells to "take in" the DNA. Sometimes the goal is to "clone" a particular piece of DNA, often by

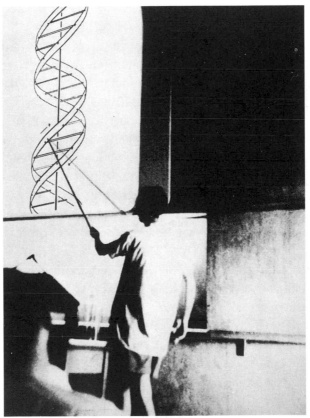

FIGURE 5

James D. Watson giving a talk on the newly published double helix structure of DNA at the Cold Spring Harbor Symposium in 1953. *Courtesy of the James D. Watson Collection, Cold Spring Harbor Laboratory Archives*

first incorporating it into a larger segment that has the ability to self-replicate (such as the genetic material, or "genome," of a virus). Sometimes, using already cloned DNA, the objective is to alter the genetic makeup of the cell. If the cell is mammalian, and able to be fitted into a nascent embryo, the grown animal or its descendants may demonstrate a predetermined genetic trait. Or similar techniques may be used to "knock out" some normal gene. The leverage that this provides in studying how particular genes affect embryogenesis, or the formation of specific tissues and organs, is clear.

The subject matter that defines molecular biology is now evident—concrete, genetically specified functions, usually associated with particular proteins, that can be enumerated and analyzed down to the level of naked DNA. Virus research has had a catalytic role in revealing basic mechanisms that underlie the infinite-seeming variety of biological phenomena. The evolutionary conservation of molecular mechanisms lies beneath the variety and supports its existence. The specificity of the mechanisms provides multiple points of entry and a degree of certainty in new research and in efforts to cure illnesses.

The existing paradigm for molecular biology rests securely, though not exclusively, on the work of Avery, Hershey and Chase, and Watson and Crick. A vital cornerstone was also provided by Edward Tatum and George Beadle (joint Nobel award, 1958). They assembled data before the middle of the century demonstrating that individual genes correspond to (nowadays we would say *encode*) individual enzymes (= proteins). "One gene, one enzyme" was a shorthand way of expressing the first clear insight into what genes *actually do,* generically, in molecular terms.

In a larger historical perspective, the beginnings of modern biology can be traced to the new spirit of curiosity in the fundamental scientific revolution that emerged in Europe during the Renaissance. The transformation in art preceded that in science. We see this in the paintings (e.g., Giorgione, Bellini) where scenes from nature as well as religious themes are depicted with a fresh eye for realistic detail. Secular and religious are symbolically united in

Leonardo's "Annunciation": set in the out-of-doors and uniting exquisite observation of natural detail with a sense of energy and religious mystery.

Order in nature and the transformations of energy fascinated scientists of the sixteenth and seventeenth centuries. The drama of Galileo's clash with the Inquisition should not divert us from seeing the tension within the scientist himself between religious faith in divine order and a newfound empiricism in investigating natural phenomena. Newton also typifies this tension. The properties of matter and its interaction with energy fired his scientific imagination and his sense of divine mission. He was a precise observer according to the principles of Francis Bacon and he coupled this with the formulation of quantitative laws amenable to mathematical treatment. Thus his physics became the model for how scientific explanation worked, at a time when physiology and microbiology (Harvey, Leeuwenhoek) were still largely observational.

Molecular biology rests fundamentally on foundations of physics and chemistry laid down over the preceding centuries. Chemistry—biochemistry, properly speaking—is the bricks and mortar of this edifice. Since Heisenberg enunciated the "uncertainty principle" of quantum mechanics there have been periodic attempts to ground some supposed fuzziness or "wiggle room" in biological explanations on this principle (freedom of the will is a favorite). These attempts, nostalgic as they may seem, ignore the fact that chemical reactions, anchored in quantum mechanics, work perfectly well, that is to say predictably, whether in a rosebud or a frog's egg. There is no particular reason, apart from wishful thinking, why cells in our brains should be any different. Roses, frogs, and fits of rage look different, are different, but so are the DNA sequences and resultant proteins that support them.

"All is fire," said Heraclitus, presciently and imprecisely, two and a half millennia ago. The obvious modern and precise equivalent is Einstein's equating of matter with energy ($E = mc^2$), but a more homely version is the oxidation (burning, if you will) of nutrients to provide energy for myriad chemical reactions in living things. For our understanding of the role of oxygen in literal burning of flammable materials we are indebted to Scheele, a Swede, and Priestley, an Englishman, who independently discovered this gas, and to the experimental ingenuity of Cavendish, also in England, and the theoretical genius of Lavoisier, who laid the foundations of modern chemistry, before his life was claimed by the French Revolution. Fifty years later the German physiologist/physicists Mayer and then Helmholtz applied the new concept of *conservation of energy* to the consumption of food and of oxygen, production of heat, movements, etc., of living creatures, to affirm that they (us included) obey this most fundamental natural principle. The groundwork for this development had been prepared by Lavoisier, who demonstrated a similar ratio of heat to carbon dioxide, either when produced by animals or by candle flames.

Molecular oxygen was the thin edge of the wedge opening the way to see ourselves as simply part of the world—but now in molecular terms. The rain forests of Brazil—the green earth worldwide—are not precious just for their food production and sheltering of countless species of life forms but as sources of oxygen, produced by photosynthesis, for all of "us." (It is interesting to remember, in this regard, that Greek and Roman adherents of Stoicism, innocent of chemistry, who did not adhere to the polytheistic religion of the state, believed that in breathing they brought into their bodies a pervasive, universal life "spirit"—hence *respiration* and *inspiration*, also *to expire*.)

Unlike other spheres of human endeavor, understanding at the molecular level of plants, animals, even ourselves, seems to create a chasm between us and the objects of our knowledge. The specialized knowledge of a gardener, an architect, or a musician does not seem to do this. Perhaps this is because effects on the quality of life seem more immediate. Or, is it that we secretly rebel against prediction and certainty and prefer to keep an element of guesswork and luck in the outcome—even leaving room for some sort of "magic"? The atoms of Lucretius, jostling in the void, seem drab—

until we read the poem and find them at the basis of all the variety, beauty, bestiality, emotion, and action of history, collective and individual. In our own time New Age systems of belief, ranging from naturopathic medicine to mysticism of all kinds, are espoused as bulwarks against a perceived tide of materialistic explanation.

Mysticism as a retreat from the engagement of reason with experience—at the breakfast table or in the laboratory—should be sharply distinguished from a sense of mystery in having a life and exercising one's intellect. As Einstein once said, "The most beautiful experience we can have is the mysterious. It is the fundamental emotion which stands at the cradle of true art and true science."*

And yet there is a persistent anxiety—captured by Yeats in his poem "The Second Coming." "Things fall apart; the centre cannot hold . . . everywhere the ceremony of innocence is drowned; . . . And what rough beast, its hour come round at last, slouches toward Bethlehem to be born?" The breathtaking progress of our science, particularly as applied to ourselves, has cut our culture away from its traditional moorings. One might say that, on the face of things, science has weakened the traditional "revealed religions" that undergird more than half the world's population. This is undoubtedly a root cause of the traditionalist fundamentalism and novel cults springing up in reaction to this trend. Any parent in our society must muse in quiet moments about the kind of human world children and grandchildren will see as history unfolds. Which familiar things, which make us feel "at home" in the world, will be missing? Will "nature" seem the same?

In the traditional picture of the alchemist—the midnight candle, the sulfurous fumes, the skull on the writing desk—we can still see the image of the medieval sorcerer. Something of this still survives, the sense of something uncanny, potentially dangerous, in our perception of the alchemist's descendant, the modern scientist. It is comparatively recently that biologists have produced this reaction,

---

* For a longer passage from this same quotation, see chapter 7.

related to the spectacular development in molecular biology and its offshoot genetic engineering. But an uneasy feeling was already there more than a century ago in the debates that raged about Charles Darwin's new theory of evolution. Listen to the great English evolutionist Thomas Henry Huxley in a lay sermon he delivered in Edinburgh on Nov. 8, 1868:

> As surely as every future grows out of past and present, so will the physiology of the future gradually extend the realm of matter and law until it is coextensive with knowledge, with feeling, and with action. The consciousness of this great truth weighs like a nightmare, I believe, upon many of the best minds of these days. They watch what they conceive to be the progress of materialism, in such fear and powerless anger as a savage feels, when during an eclipse, the great shadow creeps over the face of the sun. The advancing tide of matter threatens to drown their souls; the tightening grasp of law impedes their freedom; they are alarmed lest man's moral nature be debased by the increase of his wisdom.

It is the idea of life as a mechanism that Huxley as a scientist, firmly committed to mechanistic explanations, realized was anathema to his audience. If natural cause and effect govern our lives, what is it to be human? What is our place in a world ruled by deterministic, natural laws, if these laws extend within ourselves, even into the innermost pattern of our thoughts and feelings? This is one question with which we shall be concerned in this book.

A related question grows out of the debate begun by Darwin that still swirls around us today. How can we feel at home in a biological world in which natural selection and survival of the fittest appear to mean survival of the strongest and the luckiest? Are all the exquisite creatures of water, earth, and air really engaged in a ferocious struggle for existence; is nature really, as Tennyson put it, "red in tooth and claw"? And how about our human society with its roots in nature?

Ever since nineteenth-century apologists for capitalism used Darwin to justify the rich getting richer, simpleminded economists have leaned on natural selection to justify their doctrines. In the 1980s "Reaganomics" was a fairly recent attempt to unite a fundamentalist Darwinism with an economic theory justifying acquisition of wealth. An alternative view of nature as a harmonious, nurturing environment has inspired many biologists in the last century—Huxley's grandson, Julian Huxley, as well as Rachel Carson and Lewis Thomas spring to mind—and this view has strongly influenced the ecology movement. But how hardheaded is it as science—are they just wishful thinkers in a world where ultimately it is a matter of "fitness," and only the fittest (ultimately) survive? We shall look at this in some detail, especially in terms of the scale and complexity of structures that are involved. This is related to a third major issue, which it seems to me that modern biology thrusts upon us.

This is the snowballing feeling that our own ability to change nature is altering the very meaning of the word *natural*. If we can manipulate the chemistry of the human body, if we create new forms of agriculture through genetic engineering, if we change both the environment and the forms of life itself, can we still appeal to a nature with which we want to remain in tune? The growing human population and shrinking natural resources are other facets of the problem of human intervention in nature. The most devastating intervention—though of less immediate concern than at the height of the Cold War—would be the use of nuclear weapons on a scale that would threaten the actual existence of higher forms of life on this planet. This third area of questioning opened up by scientific developments is a broad one and obviously extends beyond the effects of biology as a science. I want to consider in particular the root question of whether biology, and especially the theory of evolution in its modern form, can provide us with what Julian Huxley called a "touchstone for ethics"—some guidance in how to make our way through the altered life circumstances, the sunlit and shadowed landscape which lies before us.

First of all, the notion of life as a mechanism. Are we really an ingenious set of wheels and pulleys—granted, of molecular dimensions? Where did this notion get started? The concept of life as a mechanism is very old and can be traced back to those cities on the west coast of Asia Minor where early Greek speculative thought first mingled with the mathematical discoveries of the ancient Egyptians and Babylonians. The concept that all things are composed of a limited number of constituents was born there—water, or fire, or a mixture of simple elements. A little later the idea of atoms was added. But modern science as we know it depends on observation and experiment as well as theory, and this began in earnest in the sixteenth and seventeenth centuries, first in Italy with Galileo and Vesalius and then in the England of Newton and Harvey.

Probably an important, unsung part was the continuous craft tradition, the building of tools, of instruments, and of machines. When the circulation of the blood was described by Harvey there already existed mechanical pumps for comparison. The eye could be compared with the new optical instruments. Clocks were important. Even today clockwork is the traditional model of a mechanical device—though perhaps computer circuits are taking over that role. Post-Renaissance advances in physical sciences led to a number of complex but speculative theories of organisms as automata, beginning in the seventeenth century with Descartes (notwithstanding his philosophy of mind-body dualism). Spinoza is less specific but also sees organisms as having a material basis. Altogether the ability of people to build devices that could perform according to instructions, to a built-in design or blueprint, made the comparison of organisms to machines easier for thinkers like Descartes or La Mettrie, the author of that amusing French book of the mid-eighteenth century, *L'homme Machine*. Machines do things. What do organisms do?

One thing they do is to process and transform energy. Here the analogy was first provided by heat-requiring engines, such as steam engines (and later electrical generators), but these energy guzzlers seem crude in comparison with what a living organism can achieve on a few hundred calories of energy. In fact, the conversion of fuel to energy in the living body is so exceedingly efficient as to excite envy in the most up-to-date car designers of Detroit or Toyota City, but the source of energy in the organism and the automobile is still ultimately the same: radiation from the sun, taken up and stored by green plants (accompanied by release of oxygen into the atmosphere) and then converted either into food or fossil fuel.

Anyone who has enjoyed an hour in front of a fireplace and has stirred the embers to make the flame leap up again from a fresh log has seen the original radiant energy of sunlight from some long past summer spring again from its chemical bonding in the wood. Our bodies do the same thing, only more subtly. The energy that powers the legs of the sprinter, the brain of the chess player, or the growth of a baby comes from only one source: the sun. The sun is our magic shield against the forces of disorder and decay. These forces have their own darker power, invoked in the Second Law of Thermodynamics, the essence of which is that everything must inevitably run down, including finally the universe as a whole. And, in fact, most things do run down: cars, machines of all kinds, buildings, structures of almost any sort, and, of course, people. Why is this? It is because anything, given a chance, goes from more order toward more disorder. There are just more *ways* for things to be disorderly than orderly, whether it's an egg dropped on a hard surface or a pack of cards. Or remember the old adage among engineers, known as Murphy's Law: If there is a way that something could break down, then eventually it will break down—that way. Perhaps the poet Lucretius said it best, two thousand years ago:

> . . . whatever growing things
> You see rejoicing, swelling out, in pride
> Ascending, as it were, the stairs of life,
> These are attracting to themselves more stuff
> Than they let go of; food and sustenance
> Come easily to the veins, and pores are kept

Tight-closed enough to stop the seep of age . . .
But for a while our gain exceeds our loss . . .
From there we go, a little at a time,
Downhill; age breaks our oak, dissolves our strength
To watery feebleness . . .
. . . so things wither, die,
Made mean by loss, by blows within, without,
Assailed, besieged, betrayed, till at long last
Food fails, and the great walls are battered in.
In just this way the ramparts of the world,
For all their might, will some day face assault,
Be stormed, collapse in ruin and in dust.

. . . all things, little by little, waste away
As time's erosion crumbles them to doom.

—Lucretius, *On the Nature of Things,* book 2, trans.
Rolfe Humphries

What is it that stands in the way of this process? It is energy from the sun, energy channeled and used in very special ways—either by human hands or by that unique set of molecules, the genes, that direct the functioning of our bodies. Since our hands and the brains that direct them and all of our tissues and organs wouldn't exist without the genes, it is ultimately the genes that stem the encroaching tide of disorder. They do this not by directly contradicting the implacable Second Law, but by accomplishing something that the physicist who contemplates that law cannot readily predict from first principles. Life, as it exists on this planet, is a cosmic surprise, an unanticipated chain of events that began from inanimate atoms some billions of years ago and continues to unfold in surprises to the present day. Living things get by, by channeling the energy of the sun into their own designs, that is, the designs of the genes, and the remarkable thing is that human beings have now succeeded in deciphering so many of the intricate molecular steps by which this occurs.

This is the subject matter of molecular biology; given enough time and blackboard chalk a literate molecular biologist can give you a pretty detailed step-by-step description of how solar energy stored in the food we eat is converted, under the guidance of genes, into the materials of our bodies and the structures that participate in our actions, physical and mental. Of course, there are plenty of gaps to be filled in, especially in the realm of neurophysiology, as well as in developmental biology, that is, the complex story of how embryos develop into adults. But the barriers seem to be mainly quantitative, not qualitative, and biologists are aided by a remarkable unifying principle, that the basic patterns of life are similar in different species. At the molecular level there is a remarkable similarity between the frog and the prince of the old fairy tale, and between princesses and peas, and even between people and yeast cells.

This has all really only become clear in the last few decades, and besides making it easier for laboratory scientists, who can attack a tough problem by selecting an alternate organism for study, it has made the conclusion unavoidable that we are so much part and parcel of the rest of nature that we have pretty much got to make our peace with this fact and think twice before assigning any real biological uniqueness at the molecular level to human beings. Is there anything wrong with this, and does it invite the kind of gloomy thoughts that Thomas Huxley referred to in his sermon? I don't think so, and I hope this book will make it clear why.

But first the question of nature "red in tooth and claw." The key concept here is survival, and competence to survive—not a predator-prey situation, or a even a fight to the death over food resources (or "living space"), if the human capacity to reason is really utilized. This is not to become a sentimentalist or deny the existence of "food chains." The greatest good for the greatest sensible number (of *species*, not only individuals) is what our newfound stewardship of evolution means. Albert Schweitzer, early in the last century, called it "Reverence for Life." He acted on this principle and founded a new hospital in a part of Africa as yet isolated from modern medical advances. In our own times organizations such as the United Nations and private relief agencies have sought to ameliorate human suffering, and there have been international efforts to prevent global warming and destruction of the atmospheric ozone layer. Lists of "endangered species" have been prepared and action taken.

The clarity and concreteness of our new (half-century-old) molecular biology, which I hope to reveal in more depth in the chapters that follow, are a triumph of human ingenuity, created on the "kitchen tables" of laboratory benches all over the world. The contemplation of these hard-won truths should yield an ultimate serenity—in the spirit of self-transcendent religious meditation the world over—(perhaps especially from that part of the world that I, in my Euro-American origins, call the East).

We shall see that a deeper "touchstone for ethics" is also revealed through such contemplation.

"Grey is all theory, and green the golden tree of life," says Mephisto in Goethe's *Faust*. No Faustian bargain has been struck here. A celebration of life is before us, if we choose to understand it.

ONE

THE GENETIC POINT OF VIEW

My parents, who met in this country, had both immigrated from Germany in the 1920s. My mother, who had a doctorate in English and American literature, came to learn more about the writers of this country, my father to find some employment consistent with his own doctoral degree in musicology (few jobs existed in Germany in the aftermath of the great inflation there). They both ended up teaching German language and literature and, having met at the University of Wisconsin in Madison, moved to Wells College, a small women's college in rural New York, in a village called Aurora, on Cayuga Lake. There they quite industriously produced a number of reading books for undergraduate students of German—all original texts in prose or poetry—especially of modern writers, but also one based on "The Young Goethe." My father's real passions were music, and then painting as well (which he had begun doing as a junior officer in the trenches of World War I, as a kind of psychological relief, I now believe). My mother developed her own course in comparative literature—English, French, German, Russian—and she grew into an enthusiastic gardener of flowers and fruits and vegetables (in our extensive backyard). Both of them taught, between them, all the basic and advanced German language and literature courses needed for students to major in German.

My older brother gravitated toward writing and painting, I toward gardening and building things, and toward music (piano).

There were family "concerts" where we all listened to old 78 rpm records of Bach, Mozart, Beethoven, or Brahms, my father eyes closed in intense concentration, I myself lying on my back on the living room rug. There were also family "art shows," where my father would bring down the latest watercolors he had been working on—or my brother's pictures (even something I might have sporadically produced) so we could all comment (praise was not de rigueur). In the category of consistent praise was my mother's cooking at mealtimes—and, I believe, both my parents' performance as teachers, judging by the number of students who elected their courses and even went on to major in German during the years of World War II.

My brother was six and I was three when the war broke out in Europe. It was not easy growing up, going to school in a small town, in a family that specialized in things German, during those years. But the pressure was chiefly internal: most of our classmates were bussed in from the surrounding countryside, from farms or even smaller towns, and had no direct acquaintance with my family. The number who would go on to college was small; only about ten in the school were born of college faculty in the 1930s. For that reason, and because our town school burned down (without human injury!) in 1944, my parents decided to send my brother and myself, in turn, to college preparatory high schools ("prep schools"). In the interim I attended school classes scattered around town—in a meeting-room of a church, in the Masonic lodge, in the old post office building.

My best friends in those days, a boy and a girl, were children of two successive teachers of physics at the college. With their help I was able to borrow some leftover laboratory equipment from the college physics laboratory so I could do things like make a small telescope. I already had a kit for wiring magnets for a "demonstration" electric motor. And a kit for putting together a very simple radio receiver. And, of course, a home chemistry kit. Those were exciting diversions from the woodworking I was used to doing in the toolshed between our house and its garden. (There was scrap wood to be had from the carpentry shop at the college.)

At Deerfield Academy I had real science classes, with laboratories, in physics, and chemistry. Also four years of high school mathematics. Now my appetite was whetted; I saw science as my calling. Here I shall pass over the other pleasures and challenges, both academic and social — and also athletic — and add only that I made new friends, including one who has turned out to be my lifelong best friend, over the past fifty years.

At Yale University I opted for a rarely-chosen major, Physics and Philosophy, together with one of my roommates, a person of incisive wit and humor, whose father was the founding editor of *Commentary* (in its more liberal days) and who introduced me to some of the pleasures of urban life, such as Italian and Chinese food. We both decided as a "career-insurance" measure, and from shared interest, to add to our studies all the extra science courses in the premedical program (biology and chemistry, including organic chemistry). This could be done while still retaining essential parts of the undergraduate curriculum, in the humanities and social sciences.

The Physics and Philosophy Program was the creation of Henry Margenau, a theoretician who wrote on pure physics as well as on the history and philosophy of the discipline. We took courses in calculus and calculus-based physics, as well as the Margenau "signature" courses in the mathematical "Foundations of Physics" and "Philosophy of Physics." Then there were other advanced physics courses, with my choices being a laboratory in key experiments of the last half century, also courses in nuclear physics and electronics. And, of course, there was more philosophy after the usual introductory course: history of philosophy was required — and I chose to study the philosophy of religion also. My acquaintance with philosophy stretched back in time to discussions in the vegetable garden about Kant and Hegel, and Sartre, with my mother. Her university experience included philosophy, for example, studying with Husserl at the University of Freiburg. In 1947 she published an article in a prominent academic journal, entitled "In Defense of German Idealism," that took issue with those who saw Hegel, Fichte and Nietzsche as having, in some sense, paved the way for the outbreak of Nazi ideology. (She also published "The Legacy of Kierkegaard" in the *New Republic* in 1955, the centenary of his death.) To return to Henry Margenau, I found his approach to observation (data) and theory in science to be direct, without frills, and sufficiently encompassing to meet the obvious philosophical challenges. He was, in truth, a humanist as well as a scientist.

Another form of "reality check," with science, was provided by my part-time job, which I was obliged to work at about two hours a day, as part of my scholarship-support package from Yale. Years one and two were at first dining room–related and then purely clerical. But years three and four were connected to research in faculty laboratories—one in biophysics and one in nuclear physics. The biophysics project was led by Richard Setlow, a professor in that (in 1955) fairly new subdiscipline. I was helping to measure the "radiation cross-section," in molecular dimensions, of a particular species of enzyme molecule subjected to radiation capable of knocking out its activity. What most interested me, though, was a different project in which Setlow was involved, in which bacteria carrying a latent virus (bacteriophage lambda) were subjected to ultraviolet radiation (UV) to activate the virus. The first aim was to determine the wavelength of UV that was most effective.

To accomplish this a quartz prism, in a large wooden framework, was used to spread out the spectrum of light from a UV source, much as a glass prism can produce a "rainbow" effect with visible light. Small volumes of bacterial cultures, in UV-transparent containers, were exposed to different parts of the UV spectrum, and the geometry at which the bacteriophage was activated and burst en masse out of the bacteria was recorded. (The term used here was *lysis* of the bacteria.) It struck me that the sample of the bacterial culture was functioning like a "biological lens" through which one could study events in a single, prototypical bacterial cell. (The target of the UV here was the bacterial DNA, in which the bacterio-phage's own DNA was temporarily housed.) This use of cultures of bacteria to study events at the level of individual molecules was impressive to someone like myself, trained primarily in physics and chemistry.

When I went on from Yale to study at Oxford University on a Rhodes Scholarship my original intention was to concentrate on physics. In the months before arriving in England, while still at Yale, I reconsidered, thinking: science laboratories are rather simi-lar, the world over, and I've seen them, whereas the special thing about Oxford is having one's own tutors in certain subjects. Why not then try more philosophy, since the philosophy of science was still a major interest for me? Accordingly I signed on for the degree program in "PPE" (Philosophy, Politics, and Economics), the only program in which one could study philosophy, short of doing "Greats," which required a knowledge of classical Greek. I expected easy going in economics, with which I already had some acquain-tance, from a course at Yale, and I looked forward to studying some modern history, which the "politics" part involved. Unfortunately, as it turned out, the Oxford philosophers were in the lock grip of the school of "language analysis"—dissecting sentences to ferret out hidden contradictions or misuse of words—the idea being that the "big" philosophical problems didn't really exist, but only verbal misunderstandings of commonplace truths. Someone wanted to study "Continental" philosophy—Kant, Hegel, French existential-ism? No way. Descartes was the only thinker from across the Chan-nel (for circa the past two thousand years) who merited some atten-tion (besides, of course, Wittgenstein—his work the precursor of analytic philosophy).

In a way this was a useful lesson, for the analytic school was also getting a grip on major departments of philosophy all over Amer-ica; so there lay the future of an academic career. I remember expressing my disappointment in Oxford's philosophy curriculum once at an academic/social gathering, in the presence of Isaiah

Berlin, who said, "Come and see me in my office sometime." I did this soon thereafter and, after hearing me out, and acknowledging the essential accuracy of my observations (on the one-sidedness of the curricular offerings, barring fluency in Greek), he observed, "Attending a university is like going to a restaurant—you must choose from whatever is on the menu—whatever seems special to that place." I am happy to report at this writing my impression that the heyday of analytic philosophy is coming to an end in the United States, having lasted more or less a generation, from 1960 to 1990.

At about this time, near the beginning of my second year at Oxford, I experienced a lucky accident. I was having lunch with some other students, who were studying chemistry and biology, and I learned that a distinguished American geneticist, George Beadle, was giving a special series of lectures in his field. So I began attending these lectures, just as Beadle was describing some recent work on bacteriophage genetics. The phage involved was not the kind that could become latent within cells but the more commonplace type that simply infected susceptible bacteria and produced many progeny in an hour or less, with the subsequent lysis of the bacteria and release of the new phage particles. Mutations could occur in the phage DNA (usually one long molecule), and there could be "crossing-over" between adjacent molecules, shifting mutations from one to the other. This could result in restoration of a "normal" molecule from two DNA molecules carrying deleterious mutations, with a probability dependent on how far apart these were on the DNA "map." Thus the frequency of such normal-DNA-carrying phage particles emerging from a deliberate coinfection of a bacterial culture with stocks of two mutant phages allowed one to calculate physical distances of the mutations along the DNA.

The same basic principle has been used to map mutations on chromosomes of many species in the past century. In "higher" organisms such as ourselves, the crossing-over can occur when chromosomes of reproductive cells are lined up in matched pairs, preceding the packaging of one of each pair as a set to be together in an egg or a sperm cell. Here again, if versions of two genes that produce distinct features in adults seem very frequently to stay together, this is evidence that the two genes are close together on the same chromosome.

To return to George Beadle. He was describing deliberate coinfections of bacteria with two phage stocks, mutant *in the same gene* (the "*r*II locus"). The huge number of progeny phage were then tested for growth in a bacterial strain where only non-*r*II-mutant phage could grow. By looking for plaque production on "lawns" of the latter strain, *single* viable phage particles could be found. Thus one could map the length of the gene (*r*II) itself (or the part between the two mutations). The scientist doing these experiments was Seymour Benzer, at Purdue University. Once again I felt that this work with cultures of bacteria (and phage) was like training a powerful lens on events at the molecular level—here involving a physical dimension—length—of a single gene. This was truly quantitative biology (as in the series of the proceedings of annual meetings at the Cold Spring Harbor Laboratory, published since the 1930s as the "Cold Spring Harbor Symposia on Quantitative Biology").

I made an appointment to talk to Beadle about possible research careers in his area of biology. Beadle listened carefully to what I told him about my academic background and enthusiasm for pursuing work in basic biology, grounded in chemistry and physics. I don't think that I stressed how this appeared to me to be an antidote to my severe disillusionment with contemporary academic philosophy (in the English-speaking world). I was thinking to myself about how much more stimulating it would be to go into the laboratory every day and perform experiments, yielding real results, on living systems, than to sit in the book stacks of a university library, posing questions to myself about the meaning of life and ending up analyzing the meaning of the word *meaning* and whether the questions were "meaningful." Ironically, Beadle may have seen me as overly eager to solve the "big questions" in biology. He gave me a book to read. It was a short novel, in popular style, about a young man who wants to do his doctoral thesis work on the organism *amoeba* and find out the "secret" of how it works, how it does all the things its living state prescribes: eating, moving about, engulfing smaller organisms, reproducing, etc. All the answers about *amoeba*. He becomes bogged down in translating all the big questions into real experiments: they all overlap too much. Meanwhile some friends of his are devising experiments about more limited aspects of organisms, that yield real results: specific questions allowing specific (quantitative) answers. I got the point—pick a discrete research question and solve it. In other words, lower your sights and proceed one step at a time in your science research career—advice I followed, as I will describe in the next chapter and in chapter 4.

One day, around this time, the news broke just before one of Beadle's lectures on microbial genetics that he and his colleague Edward Tatum had been awarded the Nobel Prize for their work leading to the "one gene, one enzyme" theory, which subsequently became a foundation stone for molecular biology. Beadle was characteristically understated, after the applause subsided, and proceeded with the day's lecture, delving further into recent research on genetics of bacteriophages and their hosts.

The term *molecular biology* came into existence around that time (1958). In truth, this new scientific discipline would have been stillborn without the new advances in genetics during the 1940s and 1950s. But genetics had an explicit, and implicit, role in biological thinking stretching back at least two centuries before. We could even reach back over ten times further than that, in the history of Western thought, to Aristotle's concept of a "formal cause," which plays itself out in the development of an organism, to reach its final "form." The obvious justification of such a concept can be seen in the apparent stability of various species of organisms—in their appearances and internal allotment and structures of organs. In the eighteenth century a Swede, Carl Linnaeus, created the systematic naming of animals, and plants, by genus and species (genus: *Homo,* species: *sapiens,* for us, for example). The growth of the new field of paleontology, beginning in earnest around this time, lent new momentum to the activity of classification—fossil bones seemed to fit into identical, or related, or widely differing groups—suggesting that Linnaean categories existed in the past, some persisting into the present, many others having "died out." Species became extinct, others seemed to spring into existence over geological time. Enter Charles Darwin (1859). I will reach into his theory of the evolution of species especially for the third of three principal building blocks. 1. Individual organisms in nature undergo some variations (source unknown); 2. certain variants succeed better in surviving and having offspring (logical); and 3. *crucially,* the variations aiding survival and self-reproduction are *inherited*. There had to be some basis for inheritance of rather precise changes.

From the modern point of view we are accustomed to expecting any complex machine to have a blueprint and to be reducible to physical principles. It appears reasonable that if a structure is to be in dynamic interaction with an environment, yet stable over time (or changing in a predetermined manner), there must be a process of checking what is happening against a master plan of some sort. There must be some way to direct the substance of nutrients for a living organism to form and re-form the cells, tissues, and organs as needed. This is most dramatically the case in the process of reproduction of the whole organism.

Something akin to the modern concept of a genetic template is to be found in the speculative writings of Georges Buffon (1707–1788). He postulated a hierarchy of "interior molds" to explain the incorporation of nutrients into the fabric of the organism, and reproduction of new individuals. His contemporary Maupertuis (1698–1759) recognized the familial inheritance of human traits such as extra fingers and proposed that such heritable variations, arising through changes in germinal "corpuscles," might help to explain the diversity of species. The theme of microscopic structures carrying biological information was recurrent; Darwin spoke of "gemmules" and his near contemporary, the great German biologist August Weissmann, who experimented with frog embryos, referred to "biophors."

In some regards the figure of Gregor Mendel (1822–1884) is more central with respect to the development of molecular biology than that of Charles Darwin. In a view of life processes as consisting of molecular interactions and transformations, the resolving power of his work on genetics, with its mathematical emphasis, was of decisive importance.

Mendel's life and work, and his lack of being recognized in his own lifetime for his seminal contributions, are so well-known that to recapitulate it again here seems unnecessary. That an Austrian monk should have laid the foundation for modern formal genetics by breeding experiments with peas and published his work in 1866, with the result that it went unnoticed for a generation, seems an irony subsiding into travesty. The ratios of inheritance of dominant and recessive traits, attributable to contributions of each parental stock, that he established in plants, could be seen as easily applicable to various animals, as work with different species expanded after the 1890s. At that time the concepts of genes and "mutations" in genes became firmly established. And there was a growing suspicion that genes resided on chromosomes, those fingerlike, microscopic bodies that became visible by the use of certain dyes when observed in cells about to undergo cell division (mitosis). Notably, chromosomes appeared to be duplicated during the latter process, so that a full set of genes could be assumed to go into both of the new cells.

How could one best study mutations and the chromosomal locations of the genes they affected? The answer lay in focusing on small organisms that reproduced in large numbers. The common fruit fly, *Drosophila melanogaster*, presented itself, a century ago. It was big enough so that in a low-power microscope one could see bodily changes suggestive of mutations (eye color was a favorite), and it had a further advantage: oversized salivary gland cell chromosomes that displayed "signature" patterns of markings, visible in the microscope during mitosis. One could breed pure (inbred) strains and see how distinctive traits appeared to follow distinct chromosomal bands (in cells fixed and stained on microscope slides) in strain-to-strain crosses. The number of offspring made the mathematics more trustworthy. We have already discussed crossing-over between related chromosomes in my reminiscing about George Beadle's lectures. In Drosophila it was possible to compare the numerical frequency of crossing-over to produce hybrid strains, involving mutations, with the frequency of seeing distinctive chromosomal markings from each original strain end up together on the same chromosome. This aided in assigning mutations, and the genes they affected, to particular chromosomes. This combination of genetic constitution and microscopically visible chromosomal structures in an organism became known as the "karyotype" of the organism, and such organisms were "eukaryotes."

The scientist most closely identified with such work in the first half of the twentieth century was Thomas Morgan, of Columbia University and the California Institute of Technology (Cal Tech). Morgan's great dream was to discover the mechanism of gene action. He contributed greatly to chromosomal genetics, but he was a generation too early for the advent of real molecular biology, as exemplified by the dissection at the laboratory bench of how genes are composed of DNA, and how DNA encodes and brings about the biosynthesis of new molecules of enzyme proteins, which catalyze chemical reactions in cells.

One matter about which Morgan and his own generation were very proficient, however, was the identification of a special class of Drosophila mutations that caused extreme changes in the body plans of flies that bore them. These were the so-called homeotic mutations, which might result in an extra pair of wings, or in antennae on the head being replaced by extra legs and feet—and there were various other such "monsters." Only in the last decades of the past century were the normal genes underlying such mutations found to have relatives throughout the animal kingdom (ourselves included). The normal genes ("HOX genes") that can be thus mutated have crucial roles to play in forming overall body structure. And our own copies of these genes have a distinct relatedness, at the chemical/molecular level, to those of fruit flies. (More on this in chapter 3.)

It was in the study of specific microorganisms, the "prokaryotes" (no microscopically visible chromosomes), that the application of genetic concepts at the molecular level began in earnest. Two reasons for this are the simple nutritional requirements of prokaryotes and their large populations, the latter permitting the observation of statistically infrequent events such as rare mutations. The experimental ingenuity of Pasteur, and of other early giants such as Lister and Koch, established an empirical tradition in microbiology that culminated in the work of Beadle and Tatum on nutritional mutants. Mutants arise, or can be induced, that are unable to synthesize essential molecular building blocks—but can utilize the latter, if these are added to their growth medium. It was Beadle's and Tatum's achievement to observe that such mutants typically lacked a single biosynthetic enzyme: hence the "one gene, one enzyme" proposal that, in a single stroke, united the still arcane idea of a gene (in the 1940s) to the stepwise network of enzyme-catalyzed chemical reactions known to be happening within cells, based on research of the previous four decades.[1]

The treatment of populations of microorganisms as a kind of "biological microscope," for genetic purposes, found further application in the work of Delbrück and Luria, and especially that of Avery and of Hershey, which implicated DNA as the genetic material of bacteria and their viruses (bacteriophages). This focus on DNA led directly to the discovery of its double-helical structure by Watson and Crick, in collaboration with Wilkins and Franklin (figure 6).

With DNA established as the bearer of genetic information and a detailed, three-dimensional view of its molecular structure available, there was speculation in the latter part of the 1950s about some form of "genetic code," relating the four bases of DNA to the twenty different amino acids that constitute proteins. Frederick Sanger demonstrated in 1953 that the protein hormone insulin (as isolated from a single species) consists of a definite, invariant sequence of

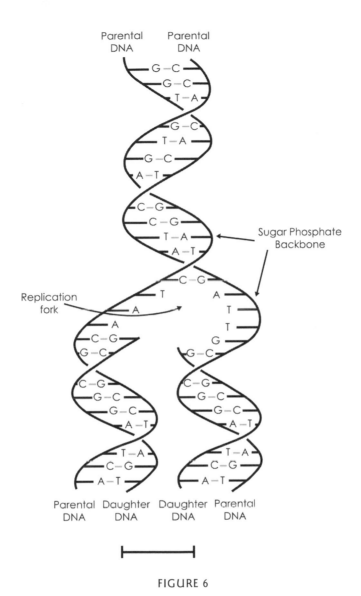

Parental DNA · Parental DNA

G — C
G — C
T — A
G — C
T — A
G — C
A — T
C — G
C — G
T — A ← Sugar Phosphate Backbone
A — T
C — G
Replication fork —— T   A
A →   T
A   T
C — G   G
G — C   G — C
C — G   C — G
G — C   G — C
G — C   G — C
A — T   A — T
T — A   T — A
C — G   C — G
A — T   A — T

Parental DNA · Daughter DNA · Daughter DNA · Parental DNA

|—————|

FIGURE 6

A diagram of a DNA double helix in the process of replicating. The enzyme active in the replication is not depicted; it would be located at the replication fork. The *bases* in the DNA, single or paired across a helix axis, are indicated only by capital letters: A, T, G, and C. For their chemical structures as base pairs, see figure 16. Alternating sugar (deoxyribose) and phosphate structures provide a "backbone" that holds together each strand of the molecule. The scale bar at the bottom of the figure represents 2 x 10^{-6} millimeter.

amino acids. The specificity of such a result had been foreshadowed by the ability of biochemists to crystallize particular proteins from purified preparations in solution, starting with Sumner's work on the enzyme urease in 1928. Four bases arranged in pairs could yield only sixteen combinations. In triplets there were sixty-four combinations, more than enough. This turned out to be the answer, with more than one triplet for each amino acid. This was the final big breakthrough for the chemical understanding of genetics: how genes, as sequences of DNA components, harbored the information for the linear structures of proteins, which could then fold up and assume their various characteristic shapes, as molecules in solution (as X-ray diffraction analysis of their crystals indicated).

In retrospect, molecular genetics developed from a complex cross-fertilization between classical genetics, microbiology, and biochemistry, and in some ways is a distinctively American hybrid. The loci were primarily East Coast and West Coast: first Morgan and then Avery in New York, then Morgan (later in his career) in Pasadena and Beadle and Tatum in Palo Alto, and, after the Second World War, the "phage group" anchored by Hershey in Cold Spring Harbor, and represented via Delbrück in Pasadena and by Luria in Indiana. (James Watson, on the threshold of his scientific career, first studied bacteriophages with Luria at the University of Indiana in Bloomington.) Delbrück and Luria, both émigrés from Europe in the context of World War II, tended to spend some time in Cold Spring Harbor in the summers, where the annual "phage course" was taught (figure 7). The exploitation of large microbial populations permitted a new level of sophistication in the selection of mutants with predefined properties, the determination of precise linkage relationships in positions on DNA, and construction of specific genetic compositions of organisms for research applications. The earlier, rather crude DNA transformation methods have been followed by newer, more powerful methods for direct transfer into prokaryotic and eukaryotic cells of defined segments of DNA. If all these developments had occurred as an extension of classical genet-

FIGURE 7

Max Delbrück and Salvador Luria enjoying a relaxed moment at the Cold Spring Harbor Laboratory in 1953. *Courtesy of the James D. Watson Collection, Cold Spring Harbor Laboratory Archives*

ics (as in Benzer's fine-structure mapping of the *r*II region in phage T4), an exceedingly precise yet intellectually limited formal system might have arisen. It was the interplay of genetics with the analysis of macromolecular structure and the enzymology of biosynthetic pathways that led to the integrated view of gene action that is now available, as we shall see in the following chapter.

We hear much discussion nowadays about human genetic intervention in biological systems—including the genetic makeup (genome) of humans themselves. The "human genome project"—defining the precise chemical structures of our full complement of genes—provoked much debate (see chapter 5). When some genes are found to correlate strongly with susceptibility to certain diseases, such as diabetes—or cancer (such mutant genes have been found for breast cancer)—the argument for their replacement by more normal versions of the same genes can be made quite strongly.

If the phrase *genetic engineering* is interpreted most broadly as human manipulation of genomes (not only human ones), by the means at hand, it becomes less problematic. It was not an idle fancy that led Charles Darwin to begin his discussion of variability among species with an account of how human beings have had their role in this. Agriculture and the domestication of animals most simply would not have happened, had it not been for the deliberate selection of certain varieties, by the choice of which ones were allowed to grow and reproduce. Seeds were saved, animal pairs were set aside as breeding stock—this was genetic engineering at its most conspicuous. Jenner's use of a milder form of pox virus, found in cows, to induce immunity to smallpox in humans (two centuries ago) was "biotechnology" too. Suppose a bovine form of AIDS virus could be used in like fashion to eradicate HIV from this planet. Would this be regarded as "low-tech," "high-tech," or simply a welcome development?

Human survival on this planet, amidst all the fossils of precursor species and descendants of the survivors of asteroid impacts and ice ages, has not been smooth sailing. Yes, we have not always made it easier on ourselves. But, from earliest times, the threats of larger animals with their own agendas, nature's tinkering with new kinds of deadly bacteria and viruses, and the vagaries of food supplies have made human innovation a key to survival. The same is true even if the risks are posed by the unreconciled passions of human nature itself, with access to our own inventions. These may range from nuclear weapons to psychotropic drugs. An intelligent, communal response is possible—as demonstrated by international treaties agreeing to dismantle nuclear weapons and banning the use of land mines. The multinational effort to devise effective drugs and vaccines against HIV is another example of humanity enlisting some of its best minds to combat the scourge of AIDS.

The risks to human life posed by disease and starvation are enormous challenges facing us in the next century. Our now half-century-old molecular biology can help in meeting these challenges all over the world. This is not simply some abstract, recondite research into the scientific basis of life. It can be the source of new ideas about how to preserve life as we know it.

The well-nigh universal pleasure experienced at seeing the young of our species—not only our own family's—start out in life is moving. It suggests to me a deeply shared faith in the future—what will be new there as well as what will be old—and in our ability to cope with it.

TWO

THE LOGIC OF THE CELL

Modern physics rests on a century-old theory that energy is exchanged and transferred from one atom to another in discrete amounts known as "quanta." In particular, radiation, whether as X-rays, ultraviolet, visible light, infrared, or radio waves, has this property, as do the energy levels of electrons within atoms and molecules. This is the "quantum theory" of Planck and Einstein and Bohr and later theorists such as Schroedinger, Heisenberg, and de Broglie.

The theory is still the underpinning of modern chemistry (though Einstein had some misgivings about its final form in his later years).

I would like to suggest that molecular biology rests on an *analogous* set of ideas, which are a radical departure from the common litany of biologists barely more than a century ago. At that earlier time the contents of living cells were generically termed "protoplasm." It was known that protoplasm contained much protein, and this was itself regarded as a chemically amorphous "colloidal" suspension. Something with the consistency of skim-milk or blood plasma, only more gelatinous. When the membranes surrounding living cells were broken, and the protoplasm released, chemical reactions were observed. These were ascribed to "ferments," enzymes, released from the cells. Eduard Büchner and his brother found that such cell-free preparations from yeast cells could faithfully perform the processes of fermentation known to be caused by living yeast. The term *enzyme* dates from that time (first years of the twentieth century) and means, literally, "in yeast."

So much for the older historical basis. After World War I there was research that showed that enzymes, and proteins in general, were discrete molecules, with definite molecular sizes and structures. Three lines of evidence, in particular, converged. Two, already mentioned in the previous chapter, were 1. the purification of proteins to the point of crystallization (the chemist's criterion of real purity) and 2. the use of such crystals to obtain patterns of diffraction of X-rays, produced by the structural repetition and regularity of the crystals themselves and indicating definite configurations of atoms in three dimensions. A number of proteins were crystallized and so analyzed in Sir William Lawrence Bragg's laboratory at Cambridge University in the 1930s.

Now, the third line of evidence. This owed much to the development of a new instrument, the so-called ultracentrifuge. Theodor Svedberg, from Sweden, had a major part in this. The basic idea was simple: spinning in a rotary device drives what is heavier toward the outer periphery, if it is free to move. Now apply this to large molecules in an aqueous medium, in a tube with its bottom pointed in that direction. Simply by virtue of their sheer size as molecules they will "sediment" toward the bottom of the tube (which is now horizontal, if the axis of the spinning is vertical). Even for large molecules, with molecular weights in the tens of thousands, the speed of rotation must be impressive to see this—tens of thousands of revolutions per minute. Obviously, as in Bragg's work, a pure solution of one protein is important here. The rest is technical design and mathematics, to observe a sharp boundary where the solution is "clearing" (of the pure protein) at the top, and calculating a molecular weight from the *rate* at which this happens. The key point is that a *sharp,* moving boundary indicates a *definite* molecular weight.

Add to all this the knowledge near the beginning of the century from German biochemists, especially Emil Fischer and his lab, that proteins consisted of a limited number of different amino acid building blocks, and, half a century later, the proof by Sanger that a sample protein, insulin, has a definite sequence of amino acids, and we have the "clincher": that proteins are not vague, gelatinous colloids but simply large molecules, with precise sizes. This is a form of *structural* "quantization" in biology (not to be confused with the *energy* quanta of quantum mechanics).

We have already discussed, in the preceding chapter and in the prologue, the present-day knowledge of the exact molecular structure of the genetic material in cells, DNA, and the genetic code by which its four constituent bases specify the precise sequences of the twenty different amino acids in the proteins that are the products of genes. This is both structural and *informational* quantization in cells. What we have not yet discussed in any detail is the *functioning* of proteins, preeminently as enzymes, catalyzing (facilitating) a vast variety of chemical reactions in cells. Immediate examples that spring to mind are the synthesis of fresh DNA and of more proteins, vital for production of additional cells and new tissues and organs, to allow growth of multicellular organisms.

This requires energy that is typically derived from food and oxygen, which is brought into organisms by breathing. Enzymes do the rest: breaking down appropriate molecules in the food so that these can be oxidized, and storing resultant energy in the form of particular "energy-rich" molecules, which can participate in biosynthesis of new DNA, and proteins, from their chemical constituents (especially bases and amino acids), which also can be newly synthesized with the participation of other enzymes, relying on the energy-rich molecules. For the enzymes involved, all of this could fairly be termed *functional* quantization. We can list the various functions and specify conditions under which they occur. We are a long way from mysterious "protoplasm."

Lest we forget: each organism—from unicellular yeast and bacteria to humans—has a *specific* number of genes, all with definite roles to play. Therefore, we must add to our overview *genetic* quantization. In chemical terms genes exhibit informational quantization; it is the gene products, proteins, in which functional quantization mainly resides. All organisms live by means of a specific and limited amount of genetic information, which is chemically defined, down to the last atom.

Part of this information is used to direct the synthesis of special components of the protein-synthesis apparatus. RNAs belong to a class of macromolecules with a strong structural resemblance to DNA: they have the same 4 bases, except that 1—abbreviated as "U"—is slightly simpler than its DNA relative—"T." Also the sugar subunit of the sugar-phosphate "backbone" of the linear macromolecule is slightly different. But RNAs can exist as relatively short "hybrid" double helices with single strands of DNA (temporarily) freed from the DNA double helix, with its base-pairing principle of organization. There are stretches of DNA in the genetic material of all cellular organisms that exist to encode RNAs for functional roles in protein synthesis. I distinguish these from the *messenger RNAs* ("mRNAs") that carry information from genes to the molecular apparatus where proteins are made. The mRNAs are also made by hybrid base-pairing; they have discrete (large) molecular weights, and can be regarded as quantized information as well. In fact, an mRNA is an archetype of such information, as it is transferred from a gene to a protein (figure 8).

Since mRNAs are made by use of the same base-pairing "language" that is found in the DNA of genes, their formation is referred to as *transcription*. The process of protein synthesis, involving formation of a precise linear sequence of amino acids—a different "language" for the genetic information—is referred to as *translation*.

In the mid-1950s Francis Crick had an insight: there could be a kind of "adaptor RNA" to bring individual amino acids to the appropriate triplets of bases to be linked together into proteins. Not long after this the appropriate synthesis enzyme for RNAs, "RNA-polymerase," was discovered in bacteria and then in mammalian cells (figure 9). And a set of small RNAs, now termed transfer-RNAs, or "tRNAs," were found, and each was subsequently shown to have a

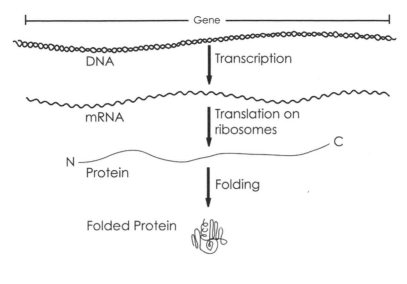

FIGURE 8

Steps in the transfer of information from DNA of a gene to a protein molecule. See text for further explanation. "N" and "C" refer to chemical groupings at the beginning and end of the protein as it is assembled on the ribosome.

recognition region for a specific base triplet in RNA (figure 10). Enzymes were found that attached individual amino acids to their own tRNA types. Submicroscopic particles called "ribosomes" were found, which attached readily to mRNAs, and to tRNAs, and with the help of some enzyme equipment completed the synthesis of new proteins (figure 11). As is obvious from this discussion, this synthesis results from bonding together of amino acids juxtaposed on adjacent base triplets of mRNA, because their particular tRNAs recognized these triplets. In effect, the ribosomes and the mRNAs move relative to each other in this process, with amino acid loaded and empty tRNAs going on and off the ribosomes. The empty tRNAs can be reloaded by their appropriate enzymes and the process continued until a triplet signifying "stop" is reached on the particular mRNA. Energy is made available from two familiar energy-rich molecules. The ribosomes have sites that accommodate tRNAs, attracting them in addition to the triplet recognition that occurs at one tip of each tRNA. This is typical of the "fit" that occurs between large molecular entities in cells. We shall discuss this subject at more length in the next chapter; I mention the tRNA case here because it relates to my own doctoral degree research and to my postdoctoral experience as well.

I chose to work for my Ph.D. degree at the Columbia University Medical School, in its Department of Biochemistry. I began in 1959. There was a strong emphasis on chemistry in the department, thanks to a group of distinguished émigré scientists from Europe. I had met the department chairman, David Rittenberg, celebrated for his pioneer work, with one such émigré, in using chemical isotope "tracers" in biological systems, before I left for Oxford. I liked his dry wit and a kind of "no-nonsense" style that seemed consonant with his generation, which matured during the Depression years in this country. One of the émigré scientists was Erwin Chargaff, a dignified and distinctive-looking Austrian biochemist. Chargaff had discovered information, by direct chemical analysis, that was essential in the deduction by Watson and Crick that DNA could exist as a double helix: that the four DNA bases could be seen as two pairs in which the amounts *within* a pair were always equal.* Hence the possibility of base-pairing between the two strands of the double-helical structure.[1]

When I returned from Oxford and entered the Columbia department, I was reading recent journal articles one day in the departmental library, when I ran across one from a member of our own faculty that interested me. Normally bacteria keep a balance between production of new RNA and new protein. If protein syn-

---

* The two pairs are A with T and G with C. Thus, in molecular-equivalent amounts, A = T and G = C. The contributions of A and T versus G and C to the overall composition of a DNA can vary depending on the biological source of the DNA. See also figure 16 before note 1, prologue.

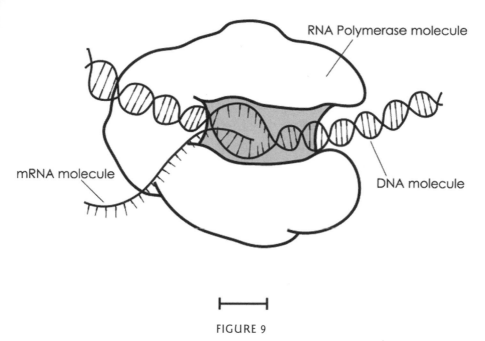

FIGURE 9

A simplified model of an RNA polymerase enzyme molecule as it catalyzes the synthesis of an mRNA from DNA. DNA and RNA bases are represented simply as short straight lines projecting from the backbones of these molecules. Under enzyme guidance the nascent mRNA strand is temporarily base-paired to one strand of the DNA, where the base-pairing of the DNA itself is transiently interrupted by the enzyme. See figure 6 for a diagram of a DNA double helix. See figure 16 for the chemistry of base-pairs applicable to DNA, RNA, and DNA-RNA hybrid strand interactions. The scale bar at the bottom of the figure represents 3×10^{-6} millimeter.

thesis shuts down, so does RNA. However, Columbia's Ernest Borek had found a bacterial mutant unable to make an essential amino acid—and thus devoid of new protein, unless the missing amino acid was supplied in its growth medium—that happily went on making RNA in the absence of the amino acid. An apostasy! Furthermore, and only very recently, Borek's lab had found that the "contraband" RNA was missing a peculiar, but normal, modification of some of the bases in tRNA: addition of a small chemical grouping to a base, termed a "methyl" group. This raised the question of how the modified base got into the tRNA in the first place, in normally growing bacteria. Was there some way to influence the base-pairing during RNA synthesis from the DNA? Or was there a later step, "after the fact," modifying some of the bases in the new RNA molecule? In talking with Borek, we both suspected the latter. This became my Ph.D. thesis problem—in full conformity with George Beadle's earlier-delivered advice to focus on a specific question. Borek himself was an émigré Hungarian, a big man with a taste for good food and for skiing, both to my taste as well.

The "missing" amino acid in the mutant bacteria was methionine, frequently itself a source of methyl groups in various cell types. It was a *second* mutation that divorced RNA synthesis from protein synthesis. The coincidence was that the first mutation was in the supply of methionine—hence the new RNA was "methyl-poor." So I decided to extract a mixture of proteins from normally growing bacteria and see whether I could use this with radioactive (^{14}C-labeled) methionine, and an energy-rich molecule, to render pure tRNA extracted from "starved" cells radioactive. This worked, and I could show that the ^{14}C was incorporated only into the rare, methylated bases. It seemed likely that there existed some specific enzyme, or enzymes, to accomplish this. The next step was to begin purifying these enzymes from the extracted bacterial protein mixture. This I was able to do and thereby demonstrate that there were distinct enzymatic activities responsible for methylating different bases in the tRNA derived from the "starved" bacteria.

A hand-lettered sign by someone was placed over the entrance to the main graduate student research lab in our department. (Some students worked in their faculty mentors' own labs; many, myself included, had a standard length of laboratory bench—a dozen feet or so—in the "big" lab and went to consult with their professors or were occasionally visited by them.) The sign read: "Lasciate Ogni Speranza, Voi Ch'entrate" (from Dante's *Inferno:* "Abandon All

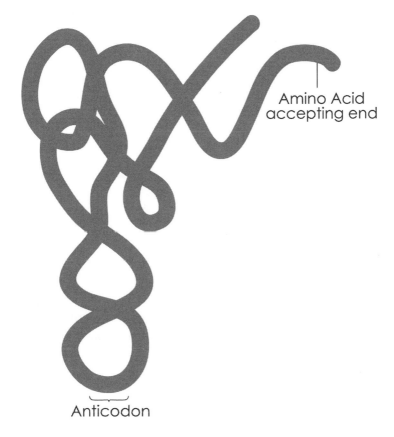

Amino Acid
accepting end

Anticodon

FIGURE 10

The overall shape that is taken by a tRNA molecule, as analyzed by X-ray diffraction. Bases projecting from the backbone of the molecule are not shown. Robert Holley received a Nobel award for tRNA base sequence data providing evidence for this structure.

Hope, Ye Who Enter Here"). This was a reference to the long hours, and not infrequent frustrations, of laboratory research. But I was having the opposite experience. I was in "Hershey Heaven," a phrase modeled on a description of an ideal research experience, by Alfred Hershey, pioneer of bacteriophage research at the Cold Spring Harbor Laboratory, the field of research for which he shared a Nobel award in 1969 with Salvador Luria and Max Delbrück. Hershey once wryly remarked that "heaven" for a researcher was to find an experiment that worked and do it over and over in different forms, to gain new information. My version was to methylate methyl-starved RNA *in vitro* ("in glass"—test tubes) with isolated enzymes and their preferred biological methyl-donor molecule, an activated version of methionine that became commercially available around this time.[2]

Writing the Ph.D. thesis was straightforward work: I had already published, with Borek, the main finding; further results were accepted for publication; illustrations had been prepared for these papers, or for national meeting presentations. The thesis was a description of the discovery of particular new enzymes involved in basic functions of living cells—biochemistry inspired by genetics in a bacterial system. Very much in the tradition of Beadle and his contemporaries.

Near the end of my period of doctoral thesis research the concept of enzyme-dependent transfer of methyl groups to RNA was extended to DNA —in order to account for the rare appearance of methylated bases in DNA. DNA-specific methylating enzymes were found by a scientist working in Borek's lab and also by a member of the laboratory of Jerard Hurwitz at the Albert Einstein College of Medicine in New York. It has turned out in subsequent years that, in eukaryotes, the enzymatic methylation of (a small proportion of) DNA bases near the point where mRNA synthesis is initiated can have a profound effect in the regulation of transcription of the gene in question. The effect is frequently to block mRNA synthesis, preventing expression of the gene. This is still an active area of research.

I decided to look further into the function of the tRNA enzymes in postdoctoral work. One lab that attracted me was that of Fritz Lipmann, a Nobelist at the Rockefeller University (the rechristened Rockefeller Institute) in New York. I wrote Lipmann a long letter detailing my interests, and he replied that he could support me with laboratory space and a postdoctoral fellowship in the fall of 1963. Soon after that we had a meeting in his office, and I could

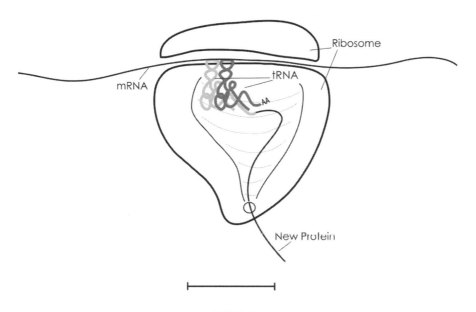

FIGURE 11

A schematic model of a ribosome engaged in protein synthesis. Two tRNAs are bound to the large ribosomal subunit, with their anticodon base triplets paired with bases of two adjacent codons of the mRNA. The amino acid carried in by one tRNA is already part of the new protein chain; the amino acid of the second tRNA (symbolized as "AA") is about to become part of the chain. This tRNA will then move (as the mRNA moves) to the position of the first (as that tRNA departs), and a new amino acid–charged tRNA will arrive, binding to the next triplet of bases on the mRNA. For clarity the tRNAs as shown are somewhat larger, relative to the ribosome, than they are in reality. The scale bar at the bottom of the figure represents 10×10^{-6} millimeter.

appreciate his informal style of getting right to the point in a conversation (figure 12).

Lipmann's lab was at the forefront in unraveling the molecular process of protein synthesis in ribosomal systems, including the contributing enzymes that were found in the soluble fraction of cell extracts. Most of the research in the lab had to do with molecular aspects of protein synthesis in cell-free preparations from different organisms. A particularly hot topic was the enzymes directly responsible for forging new bonds between amino acids on the nascent protein chain as it moved along the mRNA. Discoveries made about these enzymes became the standard of the field for years to come. One was relatively temperature-stable ("T_s"); another was unstable ("T_u"). This finding was made by Jean Lucas-Lenard, who was a postdoctoral fellow in the lab at the same time that I was. Earlier in his career Lipmann had become renowned for calling attention to the role of energy-rich molecules in biological systems. The most prominent of these, highlighted by Lipmann, was related to the "A" base of DNA and RNA, and was abbreviated as "ATP." This type of molecule was where much of the energy derived from nutrients first ended up. This energy was then consumed in carrying out various essential biosynthetic functions of the living cell: synthesis of new DNA, RNA, and protein as well as construction of other less complex molecules not so directly on the information highway from the genes to the rest of the cell—it being known that many enzymes require energy to carry out their catalytic functions.*

I have mentioned earlier something relating to the general topic of tRNAs and ribosomes, which concerns the intimate fit of amino acid–charged tRNAs to ribosomes. I found in my work in the Lipmann lab that the normal versions of tRNAs were more efficient in helping forge new proteins under the standard in vitro conditions than their unmethylated counterparts. A logical interpretation was that the methylation led to a closer fit to the pocket on the ribosome where the amino acid–charged tRNA waited for its culminating moment, when it delivered its amino acid to the new protein in formation. Then, with the movement of mRNA along the ribosome, it had to yield its position to the next charged tRNA.

---

* Interestingly, one energy-rich molecule that was found in Lipmann's lab to have a role in protein synthesis (in relation to T_s and T_u) was the G-base analog of ATP, designated as GTP.

FIGURE 12

Fritz Lipmann, Nobel laureate in 1953, before the time that he moved from Harvard to a new laboratory at the Rockefeller Institute in New York. *Courtesy of the Harvard University Archives*

This was 1965. The truly revolutionary discovery made by the use of such cell-free systems was four years earlier, when a young American scientist, Marshall Nirenberg, "broke" the genetic code, relating the 4 bases of DNA to the 20 amino acids of protein. I still remember my jaw dropping open when I heard about his talk at an international conference in Moscow (I was in the cafeteria at

Columbia). He took the simplest possible type of mRNA—a synthetic polymer with only one base, U, hence called "poly U"—and added this to a bacterial system of ribosomes and soluble factors, with appropriate energy-donor molecules to support synthesis of fresh protein. Amino acids were added, with at least one being radioactive, to monitor results. The net result? A new protein, probably never before existing on this planet: a monotone polymer of a single amino acid, phenylalanine. Almost like using a biological system to produce a piece of plastic! But the significance was earthshattering. The first unit of the genetic code for DNA to specify protein: only Us for phenylalanine, and in all likelihood a triplet of 3 Us. The floodgates were open: Nirenberg's and other labs began to test a variety of synthetic enzyme-produced RNA messengers, and even if the precise base sequences were not known, if the percent composition in terms of the 4 bases was, and it could be varied at will (of course, not always employing 4 bases at a time, but sometimes again only a single base, or 2, or 3, or 4 in various ratios), monitoring the amino acid composition(s) of the products, it was a fairly straightforward exercise in combinatorial logic to deduce candidate triplets for all 20 amino acids. Nirenberg's breakthrough research was honored with a Nobel award in 1968. The clinching argument had been supplied by Gobind Khorana, then at the University of Wisconsin, who chemically synthesized a number of messengers of *defined* sequence, where there was no doubt as to the triplets represented (Nobel, 1968).

I remember hearing a talk by Nirenberg at the New York Academy of Medicine, not long after he had returned from Russia. There was a lengthy introduction, the suspense built up, Nirenberg began his talk, accompanied by a slide presentation, and then the slide projector broke down. After what seemed an interminable delay, it was fixed, and Nirenberg continued his talk about unmasking the secrets of the genetic code. So proceeds the glory, and the fallibility, of human endeavor.[3]

The new research field in which I and a large number of other scientists were engaged, around the world, had acquired the name "molecular biology," starting in the late 1950s. At the end of that decade a new research journal, the *Journal of Molecular Biology*, was founded in England. The hallmarks of this field were the combination of genetics with biochemistry and biophysics, and a biological way of thinking that was "Darwinism" in the most contemporary sense: what were the *reasons* that certain molecules and molecular structures had been selected in evolution and withstood the test of time? One of the papers published in an early issue of the new journal was very stimulating in this way. It was from the labs of François Jacob and Jacques Monod, of the Pasteur Institute in Paris, and the lead author was a visiting American scientist, Arthur Pardee. There was considerable interest at this time in the control of total activities for particular enzymes in bacterial cells. Two levels of control were parts of the picture: *amounts* of enzyme protein per cell and levels of enzyme *reactivity* per enzyme molecule in a cell. The first had to do with gene activity in producing more or less protein product, the second with attachment of small molecules to enzyme proteins, influencing their activity as catalysts. What Pardee, Jacob, and Monod showed, and Jacob and Monod elaborated on in subsequent papers, was most relevant to the first: that a side product of an essential nutrient could bind to a "repressor" molecule, causing it to stop attaching to a DNA sequence upstream of an indispensable gene for utilizing the nutrient, thus making expression of the gene possible. And, they also showed that further downstream was an important gene promoting entry into the cell of the nutrient. Not only that, the group identified nearby on the DNA the gene encoding the repressor protein. They termed the whole compact assemblage of information on the DNA an "operon," and theorized that it represented a general model for regulating gene expression in bacteria. Jacob and Monod subsequently received the Nobel Prize for this body of work. While at Columbia, I presented the seminal paper to fellow students, and some faculty, in my "maiden speech" at our weekly graduate student seminar, and it provoked some lively discussion. Molecular biology with its genetic component was not yet standard fare in our department.

There were, in fact, three complementary approaches to proving something in the new discipline of molecular biology. One was to have evidence for a particular gene, e.g., through mutations in such a gene. The second was to identify loss of a specific cell function, such as an enzymatic activity, because of a mutation. This could be identified by the piling up of some intermediate chemical product in mutant cells, or the loss of an apparent enzyme activity in a crude whole-cell extract. The third was the actual isolation, by purification from cell extracts, of the enzyme protein itself and determination of its molecular weight, even its amino acid composition and sequence and, ultimately, its three-dimensional atomic structure in space.* So, these three: *genetic* proof of a gene, measurement of the gene-determined enzyme *activity,* and purification of the enzyme *protein* itself—provided methodologically distinct, yet converging, pathways to the truth in molecular biology. This continued to be the case for the rest of the century, though the focus shifted gradually toward more gene products—proteins—involved in regulatory roles, sometimes binding to other molecules without necessarily making or breaking covalent chemical bonds. The repressor protein in an operon was an archetype for this. Of course, proteins having principally *structural,* cell- or organism-anatomical roles have also been known for some time.

In the two summers of my years with Lipmann, 1964 and 1965, I took three-week, intensive courses on viruses at Cold Spring Harbor. One summer it was the "phage course," started in the 1940s by Max Delbrück, Salvador Luria, and Alfred Hershey, about growing and assaying bacteriophages and their mutants, a combination of lectures and laboratory experimentation. The other summer it was a

---

* The last by X-ray crystallography—nowadays one could also use data from nuclear magnetic resonance (NMR) studies of the purified protein, combined with amino acid sequence data on the protein, even aided by NMR analysis of the protein after de novo biosynthesis in vitro using amino acids labeled with appropriate isotopes.

course in growing animal cells in "tissue culture" and infecting them with various viruses, including Rous sarcoma virus, the tumor virus, originally identified at the Rockefeller Institute in its early years, 1911. (Rous virus was first found then in a tumor-bearing chicken by Peyton Rous, whom I met over the lunch table at Rockefeller in 1965. I recall his reminiscing about his role in hands-on medical research during World War I, when he was active in devising better ways to preserve blood plasma for treatment of wounded soldiers from the battlefield.) In the course we used a particular strain of the virus capable of infecting some mammalian cells in tissue culture, without production of progeny virus, the Schmidt-Ruppin strain. For experimental purposes new preparations of this virus strain could be grown in chick embryo cells, in tissue culture.

The attraction in the new research on tumor viruses was clear to me, as it was for more experienced scientists already engaged in such research in the 1960s. As in the explosive growth of biochemical research on the infectious cycle of bacteriophages in the 1950s, the attraction was the goal of *completely analyzing* a process that was directed by a limited number of viral genes. I had heard a lecture on Rous virus by a scientist visiting Columbia, near the time that I finished my doctoral research. At the end of the talk he said that "the Hanafusas (a scientist couple working at California Institute of Technology) have found that the Rous sarcoma virus is defective." This immediately evoked thoughts about how bacteriophage lambda, upon being induced from a latent state in host cells, can "steal" a segment of cell DNA, in the process forfeiting some of its own genome and becoming dependent on coinfection of its next host by a "helper" phage to reproduce itself. In this last step, however, it can carry over, or "transduce," a gene from its previous host into its new one, its DNA still remaining linked to the transduced gene, perhaps temporarily divorced from its helper virus, if this last infection was carried out with a sufficiently diluted virus preparation. Most Rous virus types, as the Hanafusas showed, required a related helper virus to reproduce in chicken cells, though not to create the "foci" of more rapidly growing tissue-culture cells indica-tive of initial infection by a single Rous viral particle. (The foci could be used to count *individual* infectious Rous viral particles, as proven by simple arithmetic on higher and higher dilutions of a viral preparation used to infect different plates, with monolayers of tissue-culture cells. The focus assay was analogous, for counting numbers of tumor viruses, to the plaque assays for counting replication-competent bacteriophage, or analogous assays for growth-competent, cell-killing animal viruses, such as polio virus or influenza virus.)* If defective Rous sarcoma virus had stolen a gene from a chicken cell back in 1911, the question naturally arose as to *what this gene might be.*[4]

To recapitulate, the advent of molecular biology, first as a research orientation, then as a formal discipline, had opened the gates to a kind of "quantum" biology (on a far different size scale than the basic phenomena of quantum physics). Quanta in biology were the genes, the individual mRNAs, and their protein products. (One could add, for completeness here, the 4 bases of DNA, the analogous 4 bases of RNA, and the 20 constituent amino acids of proteins, all related through the base triplet genetic code.) There were also carbohydrate molecules formed by chains of sugars and the lipid components (fatty acids) and phospholipids of cell mem-branes, on cell surfaces and in their interiors—all these with defined, known chemical structures. And there were the known enzymatic *activities* of specific proteins. All these specific chemical structures and chemical reactions rested on basic physics, replacing the notion, widespread at the beginning of the twentieth century, that the interior of cells was filled by that gelatinous, indecipherable material called protoplasm, marked by some microscopic organelles, such as the cell nuclei, and by the appearance during cell division of those mysterious bodies called chromosomes. Now that

---

* Plaques produced by animal viruses can be visualized by appropriate chemical staining of infected cell monolayers on tissue-culture plates.

we had proof of a whole catalogue of defined molecules in the role of protoplasm and as parts of its attendant organelles and chromosomes, how did these molecules interact to produce not only living cells capable of self-duplication but cellular constituents of specialized tissues and organs in whole organisms? What were the interactions, the signaling mechanisms, between cells and inside of cells? How did these become subverted in disease, such as cancer?

I wanted to do research on Rous sarcoma virus. I learned from a graduate student friend in Lipmann's lab, Bob Krug, that a new research group in animal viruses was being formed at the Sloan-Kettering Institute for Cancer Research, attached to a large cancer hospital, across the street from Rockefeller. The Hanafusas had moved to New York City, nearby (soon they were at Rockefeller itself), and could be an invaluable source of counsel in some of the more delicate aspects of handling the Rous virus. This thread of my story will be continued in chapter 4. I will just add here that I accepted a position at Sloan-Kettering in the spring of 1966. It happened that Peyton Rous received the Nobel Prize later that year— the longest "latent period" before a Nobel award ever recorded, I believe (fifty-five years!).

THREE

FROM CELL TO ORGANISM

The underlying theme of the previous chapter was biological *quantization*, through defined molecules. The theme of the present chapter will be molecular *interaction* and biological *cooperation*, from the molecular level to larger-scale systems. In a sense it is obvious that a molecular biologist analyzing a living system and its components must ask the question: how does this all work so well together? This ready answer is available for the complementary question: why are these *particular* molecules present? It is *because* they must work well together—enhancing survival and thus hereditary transmission. So one looks in the laboratory for efficient and useful molecular interactions. It is "bringing Darwin into the test tube," if you will, but in a rather benign spirit of looking for how the system at hand all fits and accomplishes its aims by ensembles of molecules.

Let us start, as the organism starts, with the growth and development of the embryo, "embryogenesis." Obviously, a plan is needed—what Aristotle would have called a "formal cause"—to arrive at the fully developed form of the organism. It is readily apparent that the plan must reside primarily in the genetic information that is inherited, half of it from one parent and half from the other.* We, and other "diploid" organisms, have pairs of the same

---

* This is the "genome"—or "genotype"—and its eventual expression in the mature organism is the "phenotype" of the organism.

genes as a result, just as we have pairs of chromosomes—other than the sex chromosomes in the case of males. At the level of DNA base sequence, members of our gene pairs are not necessarily *identical,* which is where being diploid has its advantage, in compensating for one not-so-good gene copy with a better version of the same gene from the other parent. (Hence, also, the age-old stigma attached to inbreeding in a family.)

Our very early embryos look remarkably similar, under the microscope, to embryos of species that are comparatively distant relatives. This can be regarded as Nature sticking to a good solution to the problem of getting things underway. The good solution is also very logical and minimalist: a fertilized egg divides a few times to form a tiny ball of cells; these then rearrange and continue dividing to form a hollow sphere of cells (also known as the "blastula"), with a concentration of cells at one point on the inside, called the "inner cell mass." In mammals the inner cell mass is the source of the fetus and the eventual newborn. The hollow sphere of cells is where tissue originates that becomes the placenta, which enables the fetus to derive life support from the maternal womb.

One fascinating area of molecular research that has opened up in the last three decades is that of cell-to-cell communication, often by protein factors, produced by cells, which can bind to protein receptors on the outer membranes of other cells. This sets in motion a kind of intracellular signaling process, proceeding from the outer membrane to the cell nucleus, where it can result in the activation of a gene, or genes (i.e., in mRNA production) that were previously quiescent. The production of new gene products is vital to embryogenesis and development of a new organism. It could also occur that a new gene product is a more general transcriptional regulator protein, binding itself just "upstream" from the coding regions of selected genes and activating mRNA production. We shall see that it is this latter process that has a key role in the overall shaping of the organism, be it fly, fish, or mammal. In fact, an insect, the fruit fly, a subject of early post-Mendelian genetic research,* was, as we have seen, highly instrumental in the discovery of a set of genes, later found to function via transcriptional regulation, genes that lay down the basic body plans for organisms with "bilateral symmetry"—left-side to right-side symmetry for some major parts.

The subject of profound significance just referred to is the role of the set of genes known as "HOX genes," mentioned previously in chapter 1, that are found in virtually all creatures having bilateral symmetry—from fruit flies to mammals. We have thirty-nine of these genes; the common fruit fly has eight. By DNA sequence analysis it can be shown that ours arose by a basic fourfold multiplication of those in insects such as the fly. There has been some extra addition and subtraction in our four sets of HOX genes, but basically they are linked together in such a way in their chromosomal locations as to suggest an original quadruplication of the set that is linked together on the insect chromosome. Why emphasize all of this? Because the HOX genes have key roles in specifying the parts of the organism, from head to tail (so to speak), and are laid out along the chromosome in the same spatial fashion in which they function in this head-to-tail geometry. (Obviously, for us, the two ends of this axis are the head and the pelvis; our legs, like our arms, are projections off this central axis.)[1]

When the concept of gene mutation was still new, about a century ago, William Bateson proposed a search for "extreme" mutations that radically affected the body plan as possible clues to developmental mechanisms. The fruit fly was ready to oblige. There were mutants that had legs in place of antennae growing out of the top of the head. (This became christened, by its mutant form, as the gene "antennapedia.") There were flies with an extra set of wings (because of mutation in a gene now known as "ultrabithorax"—the

---

* Research on the fruit fly itself ultimately showed that, for this organism, some of the earliest steps in embryogenesis, which I've just described, follow an alternate path.

mutation characterized by a duplication in the "chest" area). These are both in the HOX gene category, from later analysis. In other organisms there are HOX genes that specifically control formation of the head and development of a brain. Other genes in the family are involved in embryonic development of wings, fins (fish), and arms or legs—and similarities in basic bone structure exist in all three of these anatomical categories for vertebrates. A point to be emphasized: there is definite relatedness in the DNA base sequences of HOX genes involving such bilateral symmetry and some members of the originally discovered eight genes in the fly. It appears that by duplicating—quadruplicating—the original set, with some extra copies, there was the opportunity to achieve new results by super-imposing the effects of multiple, closely related HOX gene proteins. This was surely an evolutionary reason for the formation of the four sets of genes in ancestors of vertebrates such as ourselves (and other mammals). (There is a range of from nine to eleven genes in each set—to be compared to the eight of the fly.) Also, our heads, and their contents, are more complicated than insect heads, though the mosaic pattern of a fly's eyes is not to be belittled. At least we must concede that insects, flies included, are older on the evolutionary time scale than are mammals and appear to have acquired the genetic material supporting a bilaterally symmetric body plan first. Some special single copy organs, such as a heart, liver, spleen, and thymus—and the mammalian brain—came later, but made good use of some preexisting DNA sequence information, with new mutations and sequence multiplications. One final point to reem-phasize about HOX genes, concerning what their protein products actually do: they appear to be factors controlling transcription of other genes, by binding directly to DNA, at specified sites.

Other genes identified in flies and named by the appearances of their mutant phenotypes have counterparts in vertebrates. The names of the genes are evocative: *wingless, disheveled, frizzled, patched,* and *hedgehog* are some of them. One can imagine how these names might describe a mutant fly, or the mutant larva of a fly. (In mammals the analogue of the last is *sonic hedgehog*—harder to imagine a physical appearance to match this name, borrowed from a TV cartoon figure.) These and other exotic-sounding genes represent, in protein terms, factors that bind to cell surfaces, the receptors for such proteins, factors involved in sending signals from the cell surface to the nucleus, and transcription factors that influence mRNA production, not necessarily by direct DNA binding—themes that will be amplified further in what follows. Molecular biologists are fond of these names. They are memorable, and they convey something of the continuity of life and the developmental processes of organisms, down through evolution. There's more to it—and more to work with in explaining it—than HOX genes, despite the central role of the latter. We do have, after all, in excess of thirty thousand genes, as estimated from the human genome project.[2]

The next area of research that attracts our attention is the *growth* and *development* of the organism from embryo to adult. This requires duplication of cells and their differentiation into specific tissues and organs. It is intuitively obvious that these processes require the cooperation of numerous gene products (proteins). This is not just a matter of the basic biosynthesis of cell components; it is a matter of *regulating* this biosynthetic activity. And it is, first and foremost, a question of how the *initial decision* is made to start the process. Here, research required further development of the new field of cell-to-cell signaling and *signal transduction*. The latter term refers especially to transmission of information ("signals") from the outer membrane of a cell into the cell nucleus, where it can result in *expression* of previously quiescent genes—or the opposite, turning off the expression of genes that are actively producing mRNA for synthesis of their particular proteins. The term *transduction* also refers to the fact that, typically, there are several steps in passing a signal from the membrane to the nucleus to a particular site on the DNA of an individual chromosome (figure 13). It is not magic, it is accomplished by specific enzyme-catalyzed chemical reactions and also by molecule-to-molecule binding, with-

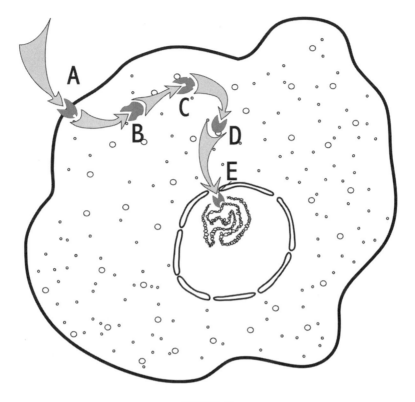

FIGURE 13

A schematic diagram of "signal transduction" from the cell membrane to DNA in the nucleus. Steps A-E involve distinct molecules and their interactions.

out necessarily forming new chemical bonds (what the chemist would term new "covalent" bonds).

This matter of molecular "recognition," without classical bond formation, is a new frontier in research. There were signs already many years ago in the discoveries of various hormones. These, it was shown, could circulate in the blood and bind to the appropriate target tissues in various organs. There were special glands discovered that produced sex hormones, adrenaline, and others. There was the "master" gland near the brain, the pituitary, that turned hormone production on and off in the others and also produced an overall growth hormone. Pituitary hormones are typically proteins that can bind to cell-membrane receptors, also proteins—a paradigm of pure molecular recognition. Recognition that has results! Secretion of a different hormone, or cell growth!

Broadly speaking, proteins involved in signal transduction belong to three major groups. First, there are those that are involved in direct binding of tissue-specific molecules to appropriate cell-surface receptor proteins, followed by the immediate transmission of the signal to the insides of cells. Second, there are proteins that affect the subsequent signal transduction process. Third, there are proteins that can themselves bind directly to chromosomal DNA, or otherwise affect the transcription mechanism for producing mRNA from particular genes.

Growth factor receptors are frequently proteins that span the cell membrane, having a part on the outside and part on the inside of the cell. When the right growth factor (usually protein) recognizes and becomes attached to the outside segment, the inside part may respond by chemically altering another protein. The alteration often involves attachment of a phosphate (grouping of oxygen atoms with phosphorus) to the other protein (figure 14). Since phosphate has a negative electrical charge, this can repel other negative charges on the surface of the protein, changing its shape as a molecule in solution. It can also enable the protein to better bind another molecule with a net positive charge on (at least a portion of) its own surface. This can induce a chain of events leading further into the cell, perhaps including other enzymatic transfers of phosphates to proteins. (The discovery that the *src* protein encoded by Rous sarcoma virus [cf. chapter 4] could catalyze "phosphorylation" of other proteins was an early stimulus for this research.)

Another class of proteins that respond to growth factor receptors are so-called G-proteins. To be active, they require the binding of a molecule designated as "GTP" (which is also a source of the "G" base found in RNA—hence *G*-protein). In their active forms

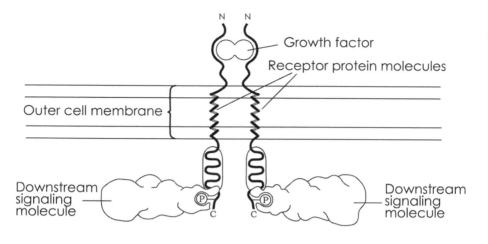

FIGURE 14

A highly schematized depiction of a pair of growth factor receptor molecules set in an outer cell membrane, where they have been drawn together by binding to a growth factor molecule. Each receptor molecule has been stimulated by this event to phosphorylate its neighbor (the phosphate groups are shown as circles with a "P" in the center). The phosphates, in turn, help to provide binding sites for two different intracellular molecules that can carry the signal(s) deeper into the cell (cf. figure 13). The scale bar represents 6×10^{-6} millimeter.

such proteins adhere to other proteins, by the molecular recognition process referred to earlier, causing the latter to alter their activity. G-proteins are frequently part of the precise molecular apparatus on the inside of cell membranes that works together with a cell-surface receptor, but they can also function further on in the signaling cascade as it leads, in precise steps, to the cell nucleus. G-proteins comprise several families of different but related molecules.

I mention this detailed information to illustrate how complex cellular processes can be dissected into discrete actions of particular molecules, acting on other molecules, in a one-on-one fashion. This is the stuff of molecular biological theory and practice—as carried out in the laboratory, and in living organisms in Nature.

We see these ideas illustrated in the third area in which I mentioned that signal transduction plays a role: the regulation of gene expression (mRNA synthesis) for particular genes. There are proteins that bind directly *to particular DNA base sequences.* (HOX gene products do this.) Often such sequences are "upstream" of where mRNA synthesis begins, clearly in a position to affect the binding of RNA synthesis enzymes to the DNA. The chief one of these enzymes, in nucleated cells, is itself receptive to binding specific proteins in a complex: protein-protein recognition again. In these cells DNA exists in chromosomes, which have specialized proteins covering the DNA, called histones. The histones must be temporarily pried loose to allow RNA synthesis. There are histone-modifying enzymes for this purpose, and *their* being directed to the sites of particular genes (to open up or close down their availability to the RNA synthesis complex) is another way that gene expression is controlled.*

When it was first announced that the human genome might contain only (?!) about thirty thousand genes, there was some hesitation in the ranks of the strict biodeterminists, who sought to explain complex human behavior on a genetic basis, and half-muffled chortles of satisfaction from the sworn opponents, who saw life as too intricate to be based primarily on genes. But, if one takes into account the various *interactions* of gene products, the combinatorial possibilities are far larger than the raw number of genes themselves. Now add the pervasive influence of family, society, education, culture—and the remembered events of a life—on human beings, and the possibilities for our bodies and brains, built on a foundation of at least thirty thousand genes (themselves subject to DNA sequence variation), seem truly immense.

---

* As mentioned earlier, in chapter 2, enzymatic methylation of DNA can also exert a strong effect on gene expression.

As we grow into fetuses, babies, and adults, certain organ systems are formed in our bodies, which involve long-range (and shorter range) signaling. A good example is the central nervous system. I want to take up discussion of this system in chapter 6.

Here let us continue with a brief discussion of how the lifetime of an organism such as a vertebrate comes finally to an end. An exploration in molecular terms of a major cause of death in humans—cancer—will be deferred until the next chapter, drawing there on my own exposure to research in the field. Since I have no intention of addressing the molecular basis of other clinical diseases and the molecular aspects of aging, or senescence, per se, are only beginning to be understood, brevity is in order.

There are repetitive base sequences that cap the ends of chromosomal DNA and grow shorter with each cell division. There is also an enzyme, known as "telomerase," that seeks to maintain the lengths of these "telomeres" but tends to lose out over time. Cells with critically short telomeres may become *unable* to reproduce—they enter the senescence phase. This is a kind of "genetically built-in" aging. There are also various kinds of tissue damage that may occur from infectious disease, or from stresses of different kinds or from the overall interaction of the organism with the environment, and require some new cell growth to restore matters back to normal. Then there are mutations in DNA of chromosomes that upset the normal regulation of genes, or damage and effectively destroy key genes, and thus block de novo production of their protein products*—in some cases as a result of breaks in the DNA. Powerful radiation or toxic chemicals can cause mutations—sometimes only changing a single base in DNA (and thus one amino acid in the protein encoded) can have a potentially negative effect. We must count our blessings that we have another gene copy in the other chromosome.

---

*   If the corresponding mRNA has some stability, protein synthesis may continue for a limited time.

There is a protein available in many cell types known as p53 that can induce "programmed cell death," known as apoptosis, and is also active in the senescence response. p53 is activated by various forms of cell stress, including DNA damage, and inappropriate cell growth (as in cancer). Apoptosis is also utilized by the genetic plan in the early development of some organisms, to shape certain tissue structures where cell growth alone would not suffice. Recent research indicates that p53 can be a powerful—genetically programmed—agent in the natural aging process of tissues and whole (mammalian) organisms.

To return to the overriding theme of the present chapter—molecular cooperation in the growth and life of a multicellular organism—there is, first of all, the capacity of molecules formed from a limited set of chemical elements to recognize each other, to form aggregates. These distinctive molecular arrays allow cellular specialization in tissues that are parts of organs in more complex organisms. This recognition process is based on geometrical shapes of molecules that "fit" each other. The fit is aided by chemical processes that do not result in the (limited set) of chemical bonds that characterize the molecules themselves. Electrical attraction of opposite charges has a large role to play here. There are less permanent kinds of bonds that are exemplified in water itself. One sees this in the geometrical patterns of ice crystals and snowflakes. Water molecules themselves are angular in shape and have an unequal distribution of charge from the hydrogens (more positive) and the oxygen (more negative). They can transiently arrange themselves in small groupings in liquid water in the same sorts of geometrical patterns (figure 15).

These groupings result from a "hydrogen bond" between a hydrogen atomic nucleus (a single proton, positively charged), on its side away from its partner oxygen in water, and the oxygen of another water molecule, on *its* side where its own pair of hydrogen atoms are more removed in the angular shape of the molecule. A nitrogen atom in a DNA base can also bind to a hydrogen in the other base of a "base

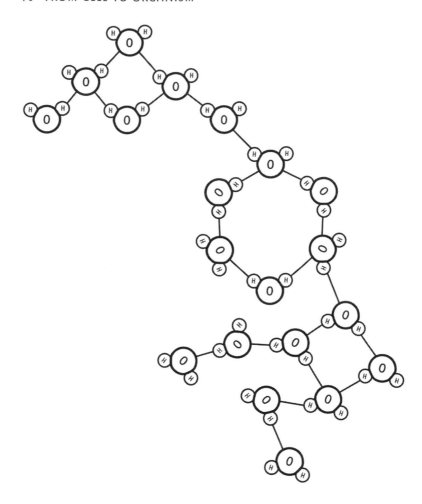

FIGURE 15

An example of how water molecules (H_2O) might together form hydrogen bonds when less disturbed by thermal motion. This is a two-dimensional picture; one can imagine extensions into three dimensions of more complex patterns of similar intermolecular linkages. Such patterns of water molecules can form around hydrocarbon side chains of protein molecules, and other such "hydrophobic" parts of large biological molecules, where thermal energy–associated movements of the water molecules are more restricted.

pair" in an analogous way. Precisely these kinds of atomic interactions bind the two DNA molecular strands in the "double helix" together and permit the faithful reproduction of new strands when DNA replicates, or is "transcribed," into messenger RNA (mRNA).*

One more example, which shows how the structuring of water itself has important consequences. There are some types of molecules that don't "get wet" easily. Examples are "oily" substances or fatty materials in living organisms (as exemplified in cooking, e.g., making soups containing meats, where fat collects and rises to the top—easy to skim off after an overnight sojourn in the refrigerator). Carbon atoms attached to one another and to hydrogen atoms are typically involved. These hydrogen atoms don't lend themselves to hydrogen bonds. Since the surrounding water can't make even temporary connections to these "hydrocarbon" structures, it falls into the well-rehearsed role of making miniature geometric arrays of its own molecules around the structures. The same occurs with hydrocarbon side chains of proteins, which are a feature of a certain subset of their amino acid constituents. But what if the hydrocarbon side chains become very close to each other? This would happen when two protein molecules recognize each other and form a close "fit." There is less space for the water to array itself around the side chains, where thermal jostling by other molecules (of water) had been less, and thus some of the geometric arrays are disrupted. This conforms to the Second Law of Thermodynamics, which states that, given the opportunity, systems move toward greater disorder. This physical principle *favors* the close association of the hydrocarbon side chains in what are termed "hydrophobic bonds"—because the *net* effect is more disorder, if surrounding water is taken into consideration. "Disorder at the service of order!"[3]

Hydrophobic bonds have important results in the fit of large molecules, such as proteins, to one another.[4] In fact, in addition to frank electrostatic attraction of negatively and positively charged side

---

* See figure 16 preceding note 1, prologue.

chains on proteins, comparable to attraction between negative charges of DNA and some positively charged proteins, there are two fundamental kinds of forces holding proteins in specific shapes and driving the fit of one to another: hydrophobic bonds and hydrogen bonds.[5]

Signaling molecules that bind to receptors on the outer surfaces of cells, and stimulate addition of phosphate groups to target proteins in the interiors of these cells, stand out because the phosphates carry negative electric charges, which are unusual alterations, facilitating further transmission ("transduction") of the signal. Metal ions, such as those of calcium and magnesium, that carry a double positive charge, can also have important roles in stabilizing negatively charged groups of atoms on proteins as well as on DNA and RNA. Singly charged ions, such as sodium and potassium (positive) and chloride (negative), also have parts to play in the electrical balance of cells—the rapid movement of impulses along nerve cell fibers is a notable example.

Owing to the asymmetric distribution of electrical charge along an axis of the water molecule itself, water is hospitable to all these electrically tinged phenomena, including, as mentioned, the occurrence of hydrophobic bonds within or between proteins. A century ago the appropriateness of water as a support medium for living systems was already recognized by some researchers, without our present-day knowledge of all the nuances. The versatility of water as a solvent for various sorts of molecules was one basis for this recognition. What was, and is, understood certainly falls under the heading of "molecular cooperation."

Consider cell-to-cell signaling in embryogenesis. The soluble molecular factors that are involved include "families" of proteins that can stimulate growth of nearby cells, or cells further removed, that have the appropriate receptor proteins on their surfaces. There are cells called "fibroblasts" that are plentiful in embryos, contributing to overall tissue mass. Their growth is stimulated by "fibroblast growth factors" (an unsurprising name for some very powerful agents on the molecular scene). When established cell lines of fibroblasts from mouse embryos are "transformed" into cells that grow in a different fashion under laboratory conditions, in tissue-culture plates, by controlled genetic alterations, they may release a protein that aids and abets the transformation process, called "transforming growth factor-beta" (TGF_β). I saw this process myself when I was engaged in cancer research. In cancers TGF_β can also function as an inhibitor of growth.* There is a family of TGF_βs, with roles that range from control of cancer cell growth to certain normal cell functions, including events in embryogenesis.

I shall mention just two other secreted protein families that have important roles in shaping our bodies—and bodies of many, many other species—including, once again, fruit flies. This body-shaping process involves the positioning and growth of cells and tissues along major body axes, such as front to back and top to bottom ("head to tail"). First there are the "Wnt" proteins, which are key players in the patterning of specific body elements and tissues to form wings, fins, or limbs. The second secreted protein family comprises the "hedgehog" proteins, which can influence cell positioning in the same body appendages.† The hedgehog signaling system also helps to determine cell fates in the developing nervous systems of vertebrates. The genetic analysis in the fly, which provided a context for studies of these two secreted protein families, and of other basic genetic commands in embryogenesis, was especially active in the laboratory of Christiane Nüsslein-Volhard in Germany, earning her share in a Nobel Prize with her collaborator, the American scientist Eric Wieschaus.‡

It is remarkable that some of the same signaling mechanisms are

---

* Joan Massagué, personal communication.

† From our earlier discussion in this chapter, it is evident that both these families also cooperate with the appropriate HOX genes in arriving at developmental pattern formation.

‡ A third Nobel laureate that same year (1995) was the eminent American geneticist Edward Lewis.

active in so many different organisms—literally, from fruit flies to humans. A testimonial to the unity of life on the planet.

The regulation of transcription—gene expression through mRNA production—is a multimolecular, coordinated set of events and can be the outcome of many signaling processes. Classical hormones, produced by endocrine glands, are an example of longer-range signals that also affect synthesis of gene products by well-defined steps. Pregnancy in mammals is made possible by synthesis of new proteins due to the action of progesterone. Female and male sex hormones, thyroid hormone—the list could go on—all affect vital bodily processes by discrete molecular steps.

Since newly fertilized ova already contained a basic kit of transcription factors, they can get on with the task of regulating transcription without further ado. As the embryo develops, new potentialities arise by the appearance of hitherto dormant transcription factors.

For example, the HOX genes share a related DNA sequence (the "homeobox") that encodes a portion of the protein products of the genes. The predicted properties of the equivalent amino acid sequence made it likely that this region of the protein could bind to DNA. By appropriate in vitro experiments it was soon shown that the proteins did bind to DNA, sometimes even to the upstream regions of *their own genes* (autoregulation) and of *other HOX genes*.

Homeobox transcription factors (HOXs) must work in concert with other regulatory proteins in controlling the rich array of genes active in embryogenesis and differentiation. What we are dealing with is a network of interacting factors that can affect each other's own expression as well as that of various other genes.

The adult organism, whether fly, fir tree, or featherless biped, is an agglomeration of cells, all preoccupied with their distinctive patterns of transcription. The need for regulatory input does not vanish when an embryo has become an adult. The maintenance of differentiated function, and of the balance of function known as homeostasis, depends on a flow of information to the regulatory regions of various genes—whether in cells of a hair follicle or in some pivotal nerve cell.

Every cell in the human organism exists in connection with what the father of modern physiology, Claude Bernard in the nineteenth century, called the "interior milieu." Just as bacteria in the soil or a protozoan in a pond are alert to environmental changes (the next dinner being of paramount importance), the cells of our bodies are responsive to a rich brew of chemical signals in which they are bathed. A generation ago the major hormone systems could be counted on the fingers of both hands; their number has been swelled by a plethora of molecules not originating in the classical endocrine system. The surfaces of cells are studded with receptors, the chemistry of which is becoming known in detail, owing to the power of molecular cloning. There are growth factors of various kinds that can stimulate mitosis or help to support differentiated functions of specific cell types. There are "insulin-like" growth factors not produced in the pancreas. There are effector substances formed in the intestines. There are "interleukins" that act on blood cells and are produced by cells inside and outside of the blood-forming system. Nerve cells signal to each other with an assortment of neurotransmitter substances. Cells adjust their immediate milieu with intricate ring compounds, the prostaglandins.

Cells respond to this buzzing blooming profusion of chemical signals through an intracellular molecular orchestra equally complex. The action begins, of course, at the cell membrane, where the incoming signals are received. Ion channels in membranes yawn open and shut. Ingenious membrane-embedded enzymes click on and off, activating further cascades of enzymes. The acid/alkaline equilibrium of the cytoplasm "breathes"; intracellular stores of calcium are released and trigger changes in protein structure and catalytic behavior. All of this orchestrated activity converges on the cell nucleus, unless the cellular answer to the incoming information is merely a temporary fix, or adjustment. If the cell's decision is to turn on a particular set of genes in response to the new information, this is arrived at by a summing up of the intracellular chorus of signals. The process might be analogized to what neurobiologists believe occurs when a single neuron decides to fire an electrical impulse along its axon by integrating the various signals arriving from other neurons. In fact, the process by which a neuron progresses from such a short-term reaction to a long-term potentiation of the effect ("memory" at the level of the single neuron) seems to involve some of the same cytoplasmic stirrings that other cells experience when the milieu outside the membrane quickens.

Accustomed as we are to human speech, we think in terms of subjects and predicates. "The bat struck the ball." "The bee pollinated the flower." "The driver accelerated the car." The notion of a system that responds coordinately, by the continuous summation of activity in all its parts, is less in tune with the traditions of Yankee ingenuity that have nurtured the successes of American science. In biological systems the great advances in understanding of the past half century appear to have resulted from the *isolation* of individual molecules and their effects within the cell. This is a powerful method for tracing strands of action and reaction in the cell, and for developing reasonable suppositions concerning the changing patterns of molecular synthesis and degradation. But the more strands are added to the pattern the less a simple subject-predicate logic seems to describe its behavior. A *system* logic is needed, that delineates simultaneous effects of many intersecting pathways of signaling, information transfer, and energy conversion. The neo-Confucian philosophers of thirteenth-century China had the idea: a system without a center, for which the system as a whole functions as the guiding "center."

What is the predictive force, the deterministic potential, of such a system logic? A negative example can illustrate that we are dealing with a system structured to produce palpable effects—sometimes chaotic effects. Heart muscle cells have an intrinsic beat, which maintains cardiac rhythm. In rare cases a disastrous arrhythmia can occur, leading to collapse of cardiac function. The phenomenon is analogous to chaotic fluctuations described in various dynamical systems, which stretch the limits of precise mathematical prediction, though the equations of state are explicitly defined. Feedback is a characteristic of such systems: the state of the system at a future time is a function of its states at times immediately preceding. This is eminently the case for living systems, where feedback controls abound—it may actually be the overlap and redundancy of some of these controls that prevent frequent occurrence of destructive oscillations. (Such oscillations are familiar from loudspeaker feedback in microphones.) Where chaos can intrude, the capacity for ordered function must also exist (how else to define chaos except by reference to a more stable succession of states?). If living cells, and their aggregates in tissues and organs, avoid chaotic disturbance most of the time, are their successive states predictable? Can development of an embryo be predicted from a detailed knowledge of the molecular state of the fertilized egg? A leading geneticist, Sydney Brenner, suggested some time ago that an understanding of embryogenesis would be equivalent to the act of reassembly of the correct molecules and supramolecular structure—not a spelling out on paper of the causes and effects.*

Embryos are products of a continuous thread of evolution, cells begetting cells, since the dawn of living systems. That is one process we shall never re-create. Yet the necessity for the presence in the embryo of all its constituents is embedded in this historical process. The information encoded in its DNA is like a manuscript revised and annotated through geological epochs. Some constituents must be historical reminiscences, perhaps still causing ripples in the ceaseless surge of the cytoplasm around the nucleus. Others may be true vestiges, inert except as signs of an earlier evolutionary solution now abandoned. We may perturb this venerable molecular assemblage

---

* Some light can be shed by comparing the process of embryogenesis and development in organisms representing differing degrees of evolutionary complexity, from worms and insects to vertebrates.

by the ingenious intervention of laboratory genetics. The perturbations will tell us something about the elasticity of the system and its chemical lineaments, but it is possible that we may never *create* such a complex system straight from the drawing board. Only by painstakingly imitating the molecular dance of Nature herself will we be initiated into its age-old mysteries.

FOUR

ABNORMAL CELL GROWTH IN CANCER

As mentioned near the end of chapter 2, the attraction of the new field of tumor virus research in the 1960s was the *limited number* of genes carried by each virus, and the possibility of carrying out a complete, quantitative analysis of the proteins encoded by these genes. A logical approach to analyzing what the Rous sarcoma virus was doing in its life cycle appeared to be to find out what proteins were encoded in its genetic material, RNA, and what they were doing. In 1966, when I arrived at the Sloan-Kettering Institute, not a great deal was known about this. In labs in different places, preparations of virus particles had been disrupted in detergent and subjected to an electrical field, to drive the proteins through a flat polymer gel in adjacent paths, in order to size them by their molecular weights. (In practice, the field is switched off midway in the process so that the proteins of various preparations are distributed through the gel; they can then be fixed in place and stained to enable the naked eye, or an appropriate optical scanner, to see the pattern.) There were some obvious "bands" that were visible and also a lot of fainter bands. Were the latter really viral in origin? The viral RNA was being isolated and "sized" around this time as well, and there was not capacity enough to encode a large number of proteins. What were the possible contributions of fragments of cell membranes that might get through the rather simple procedures used to purify virus particles from the tissue culture medium of virus-producing cells?

I decided to use a different approach: also separating viral proteins according to their molecular weights, but then identifying which were actually virus-specific by an immunological test. Disrupting virus in detergent was not good for this purpose, because the strong detergent used stuck too tenaciously to the proteins. (It was this detergent-protein complex that was actually being sized by the gel with the electrical field.) An alternative method had been recently published (not for viral proteins): to unfold proteins in a strong "denaturing" solvent and then have them flow through a column, a vertical glass tube filled with polymer beads, equilibrated with the same solvent. The solvent in this case was a concentrated solution of guanidine hydrochloride, perhaps the most potent nondetergent-containing option. The proteins would come out of the column in order of size, the largest first.

I tried this on a preparation of Rous virus, which I produced by infecting a large batch of chick embryo cells in tissue culture, and then took each of the several dozen fractions I collected as the column dripped (by using an automated "fraction-collector") and put them in separate small "dialysis bags" so that the denaturant could leak out overnight and be replaced by a physiological solution that surrounded the bags in a large container. The next step was to test each of the fractions with antibody-containing blood serum from hamsters carrying Rous virus-induced tumors. (These hamsters I had infected as newborns with a homemade lot of the Schmidt-Ruppin strain of Rous virus.) Why hamsters? Because their tumors produced no infectious virus, and yet the tumor-bearing animals were known to have antibodies that were active across a range of substrains of Rous virus from chickens (including Rous "helper viruses," which could well have been a major component of what I saw).[1] A key question was whether the viral proteins would have "renatured" when they were back in the physiological solution, i.e., assumed their native structures by folding appropriately, so that the protein surfaces representing antigens would be present. A positive antigen test with the hamster antibodies would be a vindication. If no tests were positive, the most probable explanation would be lack of renaturation.

The tests were positive; there were several distinct "peaks" when the results for the linear succession of samples were drawn as a graph of the molecular weights of the antigens present, and these were molecular weights that could easily be encoded by the full-size viral RNA. Similar results were found with preparations of rapid leukemia-inducing viruses, with their helper viruses, from chickens and with slower leukemia viruses from mice. (The former could have been yielding basically a picture of the helper viruses.) For the latter I used antibodies supplied by a neighboring lab at Sloan-Kettering, that of Lloyd J. Old and Edward A. ("Ted") Boyse, who were deeply interested in leukemia-specific antigens of mice. I collaborated directly with a talented graduate student in their group, Robert Nowinski, who also contributed an iconoclastic sense of humor. I presented the results for the Rous virus proteins at a Cold Spring Harbor meeting on tumor viruses that summer. The day after my talk I encountered Jim Watson on the way to the dining room, and he said, "It looks like things are falling into place."

I was indebted to Saburo Hanafusa and his wife, Teruko, for the progress I had been able to make in growing Rous virus in chicken embryo cell cultures. I was able later to return the favor by demonstrating to Saburo and a postdoctoral fellow in his lab the technique of complement fixation for antigen detection. In the 1980s Saburo asked me to join him in organizing an RNA tumor virus meeting at Cold Spring Harbor.

At Sloan-Kettering I collaborated closely with Lloyd Old and Ted Boyse in defining leukemia virus-related elements in the genetic constitutions of inbred strains of laboratory mice. Ted Boyse, in particular, did the genetics and the immunology; my lab did the biochemistry. Mouse strains with a high incidence of leukemia, which had been known for decades, expressed viruses that could grow in their own tissues and also cause leukemia if injected into susceptible strains with a naturally low incidence of the disease. There were also viral genomes in many mice that were inducible but could not rein-fect their own tissues and had, rather, the capacity to grow in cells of other species (being thus known as "xenotropic" viruses). Some mice expressed antigens, in cells of their immune systems, corre-sponding to only a fraction of the proteins in a complete viral genome. A favorite of Ted Boyse, in this category, was the "G_{IX} anti-gen," which corresponded to the main structural protein of the virus outer membrane but was expressed as part of the cell membrane in specific tissues of certain mice with a very low incidence of leukemia. Jwu-Sheng Tung in my lab group did much of the viral biochemistry on G_{IX} and other intriguing antigens from the Old-Boyse group; Harry Snyder also made an important contribution.

Writing a paper with Ted Boyse was an experience. He had not lost a bit of the strong British accent dating from his earlier years in England, when he helped train pilots for the RAF in World War II. He had an irrepressible wit, he could recite whole passages from Kafka by heart, and his writing style emphasized strong, declarative statements, with as little scientific jargon or other mumbo jumbo as possible. I remember a visit to Sloan-Kettering by a major national grant review committee, when Ted went to the podium to speak on his area of research, and on broader aspects of research in immunology at the institute (I was also on the agenda to speak about the field of tumor virus research). Ted gave a talk in his sig-nature style—authoritative, provocative, and laced with humor. When he was finished the chairman of the review committee looked around at his colleagues and said, "Does anyone have the courage to ask Dr. Boyse a question?"

As my work on the viral proteins developed, two papers were published in 1970 that revolutionized the field of RNA tumor viruses, one from David Baltimore's lab at MIT, the other from Howard Temin's group at the University of Wisconsin. Temin had earlier advanced an hypothesis that certain clues pointed to a DNA copy of the Rous virus genome being installed in cells, as part of the infectious process. This might account for the stability of the virus-cell association in tumors and in cells from infected tissue culture "foci." (Temin and Harry Rubin, then at the California Institute of Technology, had developed the focus assay for Rous virus, at various dilutions, in the late 1950s.) It would also help to explain why small-molecule inhibitors of DNA synthesis had been shown to block productive Rous virus infection. And, finally, it might clarify why nobody, myself included, had been able to find a complementary-strand RNA, or a double-stranded RNA, that could function as a template for producing more progeny virus genomes during infection, the way things worked in other RNA-virus infections (poliovirus, influenza, etc.).

Temin had published a paper in the mid-1960s claiming that some viral RNA, made radioactive, could enter a specific complex with DNA from infected cells, if the DNA was first rendered "single-stranded"—so that one strand could form a "hybrid" double helix with the RNA. The problem was that the radioactive signal was so weak that people had a hard time believing the result.

So Temin, and simultaneously David Baltimore, had a different idea. If Rous virus made a DNA copy of its genome, perhaps there was an enzyme *in the virus particle* that could catalyze this. To test this idea they independently did the same experiment: taking a purified virus preparation, disrupting it gently, adding the energy-rich precursors of the four bases in DNA (actually of the base-sugar-phosphate structures that are the true building blocks of the DNA — single-strand—polymer), and incubating the mixture under physiological conditions. And what did they get? Single-strand DNAs that really did form hybrid helices, under the right conditions, with viral RNA. The enzyme they found was eventually christened "reverse transcriptase," since it appeared to reverse the DNA-to-RNA flow of information in the transcription of DNA to form mRNA. This gave rise to the term *retroviruses* for this whole class of viruses. The parallel discoveries of Temin and Baltimore earned each of them Nobel Prizes in 1975. Both carried their fame lightly, Temin with a habitual, mildly ironic, Middle Western sense of humor,

Baltimore with a bit of East Coast sophistication in his always direct style of speech.

In subsequent years it was shown that, upon virus infection of a cell, first single-strand and then double-strand DNA copies of the Rous virus RNA are made, and the latter copies are able to be integrated, apparently at random, into chromosomal DNA. New viral RNA can then be made by transcription from the DNA, as though it were mRNA! So, the stage was set for a version of the bacteriophage lambda analogy to operate for a "helper-type" retrovirus. Did such a progenitor virus, at the expense of some of its own genetic information, steal the information of some important adjacent chromosomal gene (probably of different genes at other times, in various cells)? Then, did the ability of this particular robber virus to transform the *next* infected cell into a tumor cell lead, in this instance, to the isolation of the virus by Peyton Rous? (Of course, in the company of a fresh helper virus.) Again the question: what was the stolen gene?

Joan Brugge and Ray Erickson at the University of Colorado showed one route to the answer. They made a Rous virus tumor in another mammal, not a hamster, but a rabbit. With the serum from the blood of this rabbit it was possible to form an antibody-antigen complex with a particular protein from Rous virus-transformed cells, which did not appear to be a viral particle protein. It emerged that the gene for this protein corresponded to DNA, discovered with a DNA sequence probe,[2] in the laboratory of Michael Bishop and Harold Varmus, in cells transformed by Rous virus, and in predictable amounts, even in *normal* chicken cells, as well as in normal cells of other species—allowing for some evolutionary variation—including humans! The gene involved was designated *src* (for *sarc*oma), but its protein product appeared to have some function in normal cells.

This also gave a retrospective boost to the "oncogene" hypothesis.* The hypothesis had been advanced in the late 1960s by Robert Huebner, a patriarch of RNA tumor virus research, and George

---

*   Oncology is the medical specialty of cancer diagnosis and treatment.

Todaro at NIH. It stated that there were genes, widely distributed in nature, and anchored in endogenous retroviruses, that had a decisive role to play in cancer. One could postulate that analogues of such genes might have roles to play in normal cell proliferation — as in growth of the organism or in wound repair, etc. But Huebner and Todaro had no *actual* genes to point to. Now there was one, namely *src*. There were soon others as well.[3]

In the 1960s and 1970s other RNA tumor viruses were found in various species, by veterinarians, and in laboratories around the world. Upon analysis, it emerged that these also were generally replication-defective (requiring helper viruses) and had also stolen a gene from their hosts. Furthermore, the genes involved were apparently playing important roles in normal organisms, since many of them appeared to have been conserved throughout evolution — even, for some, having relatives in insects (fruit flies). Harold Varmus and Michael Bishop of the University of California in San Francisco, already mentioned, were prominent scientists in these newer developments, which tied RNA tumor virus biology so closely to normal cell biology, as recognized by the Nobel Committee. Oncogenes became important tools for probing normal cellular processes and for analyzing the behavior of tumor cells without any known virus connection (as in human oncology — retroviruses do not have a significant role in human cancer).

I cannot leave the subject of Rous sarcoma virus without a word about an enzyme activity that was found to be characteristic of the *src* protein: phosphorylation of other proteins. This has already been touched on in chapter 3; it was a discovery of Ray Erickson's lab and also of the Varmus and Bishop group. What was also striking was the demonstration by Tony Hunter and Bartholomew ("Bart") Sefton, of the Salk Institute in California, that the phosphate became attached to a particular amino acid, tyrosine, in the recipient molecule. This was a new development for a protein of cellular origin — usually one of two other amino acids was the target in protein phosphorylation. This new target turned out to be a favorite of growth factor receptors on cells and was often involved in the subsequent steps in signal transduction deeper inside of cells, leading to the nucleus and gene activation.

Retroviruses have been useful tools in cancer research by their occasionally "capturing" (or "stealing") specific oncogenes. They can also cause disease—notably leukemia—by their DNA becoming integrated into cellular DNA near a potential oncogene. Such an integration event can cause the activation of transcription of the oncogene. This is because the viral DNA has a sequence of DNA bases for activation of its own transcription, which, by a peculiarity of viral DNA synthesis, is repeated at the "far end" of the viral DNA, thus able to cause activation of some nearby cellular gene (remember: viral DNA integration into chromosomal DNA appears to occur at random sites). A brief story can illustrate this. In the 1950s a virus called "MC29" was isolated by a veterinary researcher in Yugoslavia from a particular case of a chicken with leukemia. This turned out to be a defective retrovirus, able to cause a repeat of the leukemia, but requiring a helper virus to propagate itself (just like Rous virus). An oncogene was found to be encoded in MC29 and was subsequently dubbed "*myc.*" There had long been a mystery about how retroviruses apparently lacking oncogenes could cause leukemias in chickens and in strains of inbred mice. In the early 1980s Bill Hayward, an associate of the Hanafusas's group at Rockefeller, decided to test the "nearby gene activation" hypothesis in chicken leukemias. He was able to show viral DNA integration just upstream of the cellular *myc* gene in the leukemic cells. Later Paul O'Donnell's lab at Sloan-Kettering, with some input from my own lab, demonstrated the same phenomenon in leukemic mice, in which the disease was associated with an inherited "ordinary" retrovirus that was able to infect cells of the immune system in its own host. The *myc* oncogene was again being activated by infection and fresh viral DNA integration. Similar results for virus-induced mouse leukemia were found by a laboratory in Holland.

Cancer is a molecular disease in its origins. The definition is unrestrained cell growth, which must have its molecular correlates. The precise program of cell growth in the maturation of a vertebrate organism is organized around genes and their protein products. As we have seen, oncogene proteins are also found in normal cells. It is their inappropriate expression, sometimes in mutant forms, that can cause cancerous growth. The former is usually the result of mutations in other genes that help to regulate oncogene expression. The latter—mutation in oncogenes themselves—is—simply mutation. Mutations may result from very rare copying errors in the replication of DNA during replacement or growth of body tissues, or from "routine" environmental effects—some stray bit of radiation or a contaminant in the air we breathe. Or we may be exposed to more systematic mutagenic influences, such as carcinogens in cigarette smoke, or substantial chemical pollution of the environment, or excessive amounts of sunlight without precautions (skin treatments). Aggressive cancers generally arise from single cells in which more than one, statistically rare, mutation has occurred.

One product of mutation is the carcinogenic form of an important type of oncogene known as *ras*. The name comes from *rat*-derived *sarcoma*-inducing genes, found in mouse sarcoma viruses derived in the laboratory by infection of strains of rats with leukemia viruses from mice. The viruses, as in other cases, captured oncogenes from cells of the infected rats. *ras* genes are activated to be tumorigenic by random mutation, often affecting only a single base in a particular part of their sequences. Ed Scolnick, then at NIH, discovered the *ras* protein. He had some assistance from a gifted young scientist who had earned his doctoral degree with me, Ron Ellis. Ed and I became personally acquainted when we shared a bus trip from a meeting in northern Italy back to Rome. He is one of the most candid people I know, whether in private conversation or asking a question from the audience at a scientific meeting.

There are analogues of normal *ras* proteins even in yeast cells—

plantation from a close relative. (The latter isn't foolproof, since there can still be immune rejection problems.) Another drug, Iressa, active against the epidermal growth factor (EGF) receptor (a receptor with a protein phosphorylation function), has shown promise in clinical trials for treatment of some cases of large-cell lung cancer (even at an advanced stage of the disease).

Why is protein phosphorylation activity so prominent in these examples? Partly, I suspect, because cancer cells exploit some of the normal molecular signaling mechanisms in cells. It is a "delicate balance" (with apologies to Edward Albee) in cells, and disrupting such a crucial molecular balance as that controlling cell growth and replication can be disastrous, if the disturbance persists, coming from an illegitimate source.

The methods of modern biotechnology permit an assessment of *all* the gene sequences being expressed as mRNAs in a cloned cell population (cancers are clones in this sense) as well as all the detectable proteins. Conversion of the mRNA population to a matching set of DNA sequences can be accomplished by the use of purified viral reverse transcriptase. If necessary, these DNA sequences can then be cloned in bacteria by standard procedures, thus creating a DNA "library," which is inexhaustible for further analysis. These collections of molecules, literally dispensed as "microarrays" in geometric patterns of tiny spots, fixed onto glass microscope slides, can be analyzed by standard molecular-biological procedures, the results summarized as computerized information. Comparisons can then be made between cells of normal body tissues and cancer cells. Such microarray technology is very powerful and is making the "test tube on a kitchen table/lab bench" style of molecular biology seem a bit antiquated. It should lead the way in a systematic search for what is different—what has "gone wrong"—in a cancer cell. Of course, all the previous research has provided the context for interpreting "going right" and "going wrong." And more traditional lab bench approaches will continue to be the way to test some hypotheses. Overall, new drug designs for cancer treatment should profit.

FIVE

HUMAN GENOMES AND HUMAN BEINGS

Not long ago I was in a concert hall listening to the great Schubert string quintet. It is a piece of music that Schubert wrote near the time of his death and in all probability never heard performed. It is music of such transcendent beauty that, hearing it, one feels that he would be considered a composer of the first rank even if he had never written another note.

What are we to make of this gift of beauty left for us, now well over a century and a half later? Like all such gifts it is surprising and unexpected. And the same, I would submit, is true of the gifts of understanding that the inspiration of scientists and other scholars place before us.

As a biologist I marvel at the human capacity for invention and insight that reveals to us how simple and complex organisms are formed and are ceaselessly reinvented in Nature. In a world that began as an explosion of energy some fifteen billion years ago, the sheer existence of organic structures is a constant astonishment. Why, from this symmetrical, uniform fireball, such complexity?

I walk across the park in the morning and see a squirrel—or even a rat—and think: Why these elaborate forms, these exquisitely adapted structures, when there might be only nothing? And human beings *thinking* about these things—out of the quantum waves of atoms—how surprising and unexpected! And also composing and listening to music?

My former colleague at the Sloan-Kettering Institute for Cancer Research, Lewis Thomas, once wrote:

We are, perhaps uniquely among the Earth's creatures, the worrying animal. We worry away our lives, fearing the future, discontent with the present, unable to take in the idea of dying, unable to sit still. We deserve a better press, in my view. We have always had a strong hunch about our origin, which does us credit; from the oldest language we know, the Indo-European tongue, we took the word for earth—*Dhghem*— and turned it into "humus" and "human"; "humble" too, which does us more credit. We are by all odds the most persistently and obsessively social of all species, more dependent on each other than the famous social insects, and really, when you look at us, infinitely more imaginative and deft at social living. We are good at this; it is the way we have built all our cultures and the literature of our civilizations. We have high expectations and set high standards for our social behavior, and when we fail at it and endanger the species—as we have done several times in this century—the strongest words we can find to condemn ourselves and our behavior are the telling words "inhuman" and "inhumane."

(The Medusa and the Snail)

In a sense it was inevitable that, once the double-helical chemical structure of DNA was established, there would some day be an effort to describe in precise molecular terms all the genes of the human organism, along with their spatial relationship to each other in the set of chromosomes that represent "one copy" of what is in our cells—our genome. Since the genes contain the basic instructions for what happens in the cells—for the molecules of which they are composed—the implications for fundamental biology, and for health, disease, and therapy, are clear. Some genes associated with hereditary human diseases are known. Two human genes with some relationship to breast cancer have been identified, and intense searches are under way for other genes relevant to cancer, as well as for genes associated with mental illnesses.

Why then the debate that occurred about the Human Genome Project, a consortial activity of the Department of Energy and the National Institutes of Health? Why the skepticism?

Contemplate for a moment the human proclivity to make evaluative comparisons and to base real-life decisions—with implications for jobs, insurance policies, and such—on judgments of fellow human beings. The strength of public desire to find an "objective" basis for racial and ethnic prejudices has come to no final test. One quasi-scholarly book, *The Bell Curve,* claiming a classical-Mendelian genetic basis for linking race to differences in outcomes for "intelligence tests," school performance, etc., has sold a million copies. Would you want your genome or your kids' in the public domain? Also, vast DNA sequences, almost by their nature, must be stored in computers, and there are concerns about the confidentiality of databases that might prove useful to public health officials—side by side with atavistic fears about the strength of our genetic "stock" as a nation.

A critic might also ask whether the Human Genome Project is somehow inherently "antidiversity." When *one* human genome sequence is initially adopted for medical research purposes, how will the diversity get "filled in" in the databanks? If preconception or prebirth human genetic engineering proceeds apace, will some banal "eugenic" human traits win out?

Jon Beckwith and David Botstein, both eminent geneticists, have called attention to the gaps in our knowledge of the genetics of human behavior.[1] They point out that the vital influence of environment cannot be ignored and that mere correlations do not constitute cause and effect. Does darker skin "cause" a lower grade on a Scholastic Aptitude Test (SAT)? Who has factored in the effects of the SAT "prep courses" attended by kids from well-connected, well-heeled middle-class families? The observable product of the genome, or genotype, is the phenotype. As Botstein points out, there is a problem in defining human behavioral phenotypes, especially when a change in environment can cause a large shift in the data. Botstein on *The Bell Curve:* "nonsense . . . not to be taken seriously." Beckwith: "racialist science." There can be an assumption that some sort of crisp genetics is operating where there is no such thing.

The theme of human diversity has been addressed directly by Luca Cavalli-Sforza, a well-known human geneticist. His basic argument is illustrated by the following: "If you take differences between two random individuals of the same population, they are about 85 percent of the differences you would find if you take two individuals at random from the whole world." The mammalian genome is constantly reassorting itself and recombining during reproduction. This has led some thinkers, notably Richard Dawkins (in *The Selfish Gene*), to see single genes (not organisms or species) as the real players in evolution. One does not have to subscribe wholeheartedly to this theory to recognize that even in small groups of people there are enormous differences among individuals, sometimes in traits that are not immediately obvious. Traits that are obvious, such as skin color and body contours, represent adaptation (originally) to different climates. In fact, Cavalli-Sforza says, 90 percent of these "external phenotypes" correlate with a single environmental measure: average temperature. The rest of human genetic variation among individuals is the same worldwide.

Thus the concept of race has no biological foundation, except for external characteristics, which in modern society no longer have much significance for survival "climatologically" (*socially* is a different matter).

There is a consensus that the Human Genome Project can lead to advances in medicine by pinpointing genes that are involved in particular diseases with respect to causation, symptomatology, and therapeutic modalities. For example, it is well established that cancer results from mutations, usually acquired "somatically" during a person's lifetime but also sometimes inherited—perhaps in the gene copy on only one chromosome, but setting the stage for a second somatic event. The inherited conditions have been invaluable for identifying genes that have a role in particular cancers—as well as genes acting to suppress abnormal growth.

Enter the advocates for the insurance industry. Underwriters wish to know about any risks in advance and then charge accordingly. The argument goes that insurance is a zero sum game and that, if the medically disadvantaged collect money, somebody has to foot the bill. There is a direct rejoinder available here. Before the discovery of BRCA1 (one of the genes associated with breast cancer), premiums for health insurance (including coverage of treatment for breast cancer) were set at a certain level. Why not the same level post-BRCA1? If the mutated gene carriers are *charged more,* or *denied coverage,* then premium costs for the rest of the population will drop (or insurance companies will make bigger profits). When BRCA1-based research results in better treatment, or a cure, or prevention of the cancer, then everyone benefits—including those who would suffer from inherited mutations in BRCA1—and everyone's premiums can drop. What do we mean in practice when we quote the Founding Fathers as saying that we are all "created equal"? There is public acceptance of the need to provide special facilities for the handicapped in our nation, which indeed costs money. It is part of living in a civilized society.

This is roughly the position taken by Thomas Murray, a member of an NIH Task Force on Genetic Information and Insurance. Murray has described two principal conclusions of the group: that there should be no individual underwriting in health insurance based on genetics and that genetics should not be a factor in who gets health care. That is to say, where health care is concerned, need is the underlying moral principle. At a practical level, for sufficiently large group insurance plans, random individual variations are absorbed into collective risk.

Naturally occurring gene sequences, especially human ones, should not be patentable. Patents recognize and reward inventors.

But nature has invented gene sequences. A novel use for the sequence or its protein product? No problem. But patenting the *sequence* of a naturally occurring gene would be like having a patent on the fact that *Homo sapiens* has four limbs instead of six.

Obviously, the first complete human DNA genome will be followed by others, representing our fellow human beings from around the world. The basic science thrust, as well as the need to give various groups within the human species their appropriate weight, will dictate this. In a sense the Human Genome Project, as it is propagated, will be "the great leveler," revealing the shared genetic limits and variety of our shared humanity. Ours is a "species genome," which owes its historic resilience to its combinatorial potential. Far from being a platform for racism, the Human Genome Project will point to an *answer to* and a *refutation of* genetically based racism—as our common humanity is represented in all of our genes with their various alleles, passed down through generations, dispersed across oceans and continents.

SIX

ON CONSCIOUSNESS

Consciousness is supported by the nervous system. Here the primary cells are "neurons." (There is also an important cell type known as glial cells that support neurons in various ways—for example, by producing the white matter that ensheathes nerve fibers but also by helping to control nerve regeneration.) Neurons have long filamentous projections that are called "axons" that convey electrically based signals to other neurons or to target cells, such as muscle cells. They may also have another type of filamentous process known as "dendrites" that function as recipients of axonal signals.

The point of contact between an axon and another neuron is a "synapse." A signal crosses this point through the release of neurotransmitter molecules that find specific receptor proteins on the membrane of the cell being stimulated or inhibited. (An example of a neurotransmitter molecule is dopamine, a shortage of which is involved in Parkinson's disease.) The receptor proteins span the membrane, and the part inside the cell conveys the signal further into the cell's molecular apparatus, as needed. In a neuron this apparatus can cause the opening of ion channels in the membrane that allow an influx of sodium ions, altering the overall electrical balance and changing the voltage difference across the cell membrane. The local alteration in voltage difference causes a wave of channel opening to sweep along the membrane. This is the electrical aspect of signal propagation along nerve fibers (axons or dendrites). Though it involves voltage differences, it is an electrochemical process, not the same as the movement of electrons in a fabricated electrical device—lightbulb, electric motor, or computer chip—activated by voltage.

Neurotransmitter molecules are not necessarily stimulatory for the recipient cell. One example is a molecule with the acronym GABA that inhibits signal production by the recipient neuron. Such inhibitory connections increase the combinatorial possibilities in signal elaboration and transfer by a cluster of neurons. They also have potential roles in preventing overstimulation eventuating, in the extreme case, in epileptic seizures. Neurotransmitters do not have to be constantly made afresh by neurons. The leftover molecules in a synaptic junction can be reabsorbed into the neuron and used over again. (To cite Parkinson's disease again, this explains why, when a dose of medically supplied L-DOPA is converted to dopamine by an enzyme in the brain, this neurotransmitter can be taken up by the appropriate cells, with the right receptors for this absorption process—quite different from synaptic receptors.)

I have gone into some detail here to emphasize how much molecular dissection of the cellular events in neurons has been possible, and thus the firm foundation of molecular cell biology on which current knowledge rests. This has found important applications in drug development for various medical conditions including psychiatric disorders, epilepsy, pain management, and others. It is also germane to what I wish to talk about in the rest of this chapter.

The nervous systems of vertebrates, from reptiles to mammals, can be divided into a central nervous system and a peripheral nervous system. The former includes the brain and the spinal cord, and the latter the nerve connections that radiate throughout the organism and make connections with muscles and various other tissues. (The central nervous system is also involved in direct sensory perception and in initiation of specific bodily movements.) It is sensible, and generally agreed on, that mental processes such as visualization of remembered scenes or recollection of sounds, as well as imaginative thoughts about possible future events and actions, take place in the brain. In fact, modern brain-imaging techniques, such as PET scanning and fMRI, establish that such mental activities cause detectable, temporary changes in brain states.*

I would like now to advance the following hypothesis. It is *Consciousness: The Continuum Hypothesis*. It states that consciousness represents a continuum over evolutionary time, at least since the advent of central nervous systems. What has changed over this time period is the complexity and intensity of consciousness; there was a more and more attenuated version of what we experience, the further back one reaches in evolutionary time. To claim the opposite, that *all* forms of consciousness are characteristic only of ourselves or our immediate ancestors among Homo sapiens, is a form of "creationism" similar to insisting on a separate creation of human life, unconnected to the rest of the biological world. It also raises the question: at what precise moment did consciousness spring into being in our ancestors—50,000 years ago, 500,000, or 1,000,000 years—when? Did it happen first in one individual? How much more sensible to argue that a more and more attenuated version of our full-blown mental states existed, stretching back through evolution and including a variety of species also endowed with central nervous systems. Not systems as large and intricate as our own, but nevertheless systems of neurons assembled according to the same basic subdivisions of parts—forebrain, hindbrain, visual system, auditory, etc. Certainly the other basic senses of smell, taste, and touch existed—and still exist—in other vertebrates.

---

* PET stands for "positron emission tomography," and involves introduction of an isotope-labeled normal physiological molecule into the blood, so that its radioactive signal, highly localized in space, can be used to measure changes in blood circulation in a defined region of the brain. fMRI, or "functional magnetic resonance imaging," also measures blood flow by analysis according to a different physical principle. Both procedures rely on the capacity of the brain to modulate blood flow to reflect localized activity of neurons, and present to the neuroscientist-observer a detailed geometrical map of neuronal activity in specific areas of the brain.

Nicholas Humphrey, an insightful British theoretical psychologist, sees consciousness, whether in humans or in other animals, as closely related to sensory experiences and the ways in which these are processed by the brain. As he puts it, paraphrasing Descartes: "*Sentio, ergo sum*—I feel, therefore I am." Thus, he asserts, mental imagery and other conscious mental activities involve some "'reminders' of sensation."

Concentrating for a moment on other mammals, such as dogs, or horses, it is evident that they have the capacity for memory—visual, auditory, and olfactory (they remember us!). It is reasonable to conclude that their memories include recent and not-so-recent events, especially those associated with pleasure or discomfort. (Their reactions certainly suggest such emotional experiences.) There is an obvious Darwinian advantage in animals having stored up memories of certain situations and events that seemed to them beneficial or harmful. Should they seek out such situations again, to their advantage, or avoid them—to their detriment, if they fail to do so? Obviously wild animals must pick which experiences to repeat, and which to stay away from. They must recognize their mates! And where the food supplies are. Even bees remember the latter and lead their hive companions to the rich supplies of flowers.

But, you say, we humans have language, a rich vocabulary for storing ideas and communicating them—and words for "I" and "me." We have oral traditions, books, libraries, and the Web. However, this discussion is not about whether or not our mental resources and adjunct resources are more complex than those of other animals. This is about the "either-or" question of whether we have *everything* in terms of an inner mental life, and they have *nothing*. After all, animals do communicate things, by sounds or gestures. They respond to some of our spoken words. When we feel something, is the word first? Or the feeling itself?

This is why I advocate the idea of a *continuum,* over the stages of evolution leading to our grammatically informed language, and our larger forebrains. It seems appropriate, having mentioned our human cerebral cortices, to add a few words about the neural basis of sensory experience. First, the visual cortex, where what we see is registered after the image is formed on the retinas of our eyes. The visual cortex is at the rear of the forebrain and has a fine structure of neuronal "columns" that create an encoded version of the image formed by the lenses of the eyes. From there the visual information is passed on to other specific cortical areas. The same basic system exists in many mammals (the experiments on visual neuronal columns were originally done in monkeys by Hubel and Wiesel). There is also an auditory cortex, a region specializing in taste, an elaborate system for recognizing smells, and a neuronal system in the brain for mapping touch sensations onto a spatial symbol-equivalent of our entire bodies. None of this, of course, is characteristic only of ourselves. A particular experiment done on human volunteers was to use the modern brain-imaging techniques, already referred to, to see what areas of the brain were active when a subject was asked to imagine a visual scene. The result was most interesting. There was early activity in the primary visual cortex—the "first stop" after the retina. No wonder the imagined scene might seem "real." (And no wonder that in pathological conditions hallucinations are compelling—like seeing the "real thing." A more powerful version of the primary visual cortical stimulation exhibited by the voluntary human subject.) We cannot make the same explicit suggestion, to imagine a visual scene, to an animal subject. However, since it seems incontrovertible that animals have the experiences of visual memory, it would be "species-prejudiced" to deny them a cerebral mechanism that seems so basic in calling to mind a visual impression that is not staring them in the eyes. There is a continuum of anatomical development in the central nervous system as evolution proceeds. Little evidence of saltation—"jumps"—to something utterly new. It is reasonable to believe that consciousness, as a characteristic of the central nervous system, evolved in the same fashion, by degrees.

Having made this point, it is nevertheless possible to consider our human consciousness as having some distinctive, and fascinating, features. We reflect on our personal lives and our future destinies, and talk about them, in ways that it is hard to imagine other animals doing. We have a sense of personal *identity* that is essential for all of this. This doesn't seem so developed in a three-month-old or a six-month-old baby. Anyone who has had the experience of parenting a young child has seen him or her develop and has had a role in encouraging the development. Teaching the very first words that help create a feeling of *selfhood: you, me,* etc., and connecting them to gestures about place, movement, and physical bonding (holding a hand, picking a child up) are vital. And, of course, names: the child's name, "Mom," "Dad," and some others. Other words follow that refer to classes or categories of things (favorite foods and drinks, toys, animals, trees, flowers, and so on). I shall add, for brevity, only some words for physical states and perceptions: hungry, tired, sleepy, hot, cold, sweet, bright, dark.

We have here the rudiments of what the evolutionary anthropologist and neuroscientist Terrence Deacon has termed "symbolic representation," a capacity he sees as resulting from the coevolution of human language and brain capabilities.* Symbolic representation, of course, extends far beyond the simple examples that I have borrowed from child rearing. It is the essential component in abstract reasoning and conversation, or introspection, about human relationships— friendship, love, marriage, group loyalty, and "belonging"—as well as about matters we might term intellectual. Deacon sees a potential origin in early hominid-human societies for this mode of language, in social rituals and group activities. The latter include toolmaking and group hunting, especially as meat became more and more important. The former include pair-bonding and marriage rituals driven by

---

* *Speech capabilities* also deserve mention as a facet of language development in evolution. The complex sounds that we make in talking are possible because of the coevolution with the brain of the human speech apparatus, including the larynx in its distinct anatomical location.

the increased strength of social organization and by the acceptance of monogamy, to improve child care and offset the more primitive genetic rewards for males who mated as often as possible. There were additionally peacemaking rituals between rival groups that also deserved some form of symbolic reference.

Placing this in the context of our earlier discussion, of a human individual passing through different stages of life from childhood to adulthood, a key concept in symbolic representation is that of "oneself" and being aware of one's own life. Daniel Schacter, a psychologist at Harvard University, has emphasized the vital role played by personal memories in this process. A construction and reconstruction of one's sense of oneself is based on a personal history—including the "stories" one might recount to others about oneself. The role of language is again obvious, in concrete, specific allusion to details as well as in symbolic references where needed. There is a background awareness of one's own *location* in time, as well as in space, reaching back to childhood.

Stepping back for a moment from the autobiographical to a more general view of the living world, and of animals endowed with central nervous systems, I suggest that simple *location* is a potentially powerful idea in considering consciousness itself. The brain of each creature has a distinct location in space and time. This is a prima facie reason that sensory experiences, memories, and inner awareness of current thought processes form a unity, anchored in the particular organism. It is at least a necessary condition for such unity, as the organism, with its central nervous system, moves through space and time. Our human condition adds to this how parents call us by name and tell us in innumerable ways about where we stand in the world, and also give us words for our experiences. Our consciousness is made richer and more complex, but its foundation is being *here, now.*

Broadening this to specific ideas of "selfhood," some researchers have posed the question of whether other animals, even our closest relatives among the apes, can recognize themselves in mirrors. As described by Ian Tattersall, the renowned scholar of primate and human evolution at the American Museum of Natural History, chimpanzees and orangutans can learn to recognize themselves in mirrors. If they are then made unconscious and spots of paint are put on their faces, they will afterward use mirrors as aids in picking the spots off their faces. Gorillas, in general, did not do this, nor did monkey species.

The philosopher David Chalmers, at the University of California, Santa Cruz, has advanced a theory of consciousness that argues that the special qualities of conscious experience can be recognized without divorcing these completely from the world of physics. He calls this theory "naturalistic dualism." It holds that consciousness is a fact of nature alongside other facts of physical objects, such as mass, velocity, electromagnetism, gravity, and subatomic forces. It is not *derivable* from these other facts, any more than electric charge is a direct result of mass, but it is a natural fact that exists in a web of interactions with these other facts, as they do with each other. This addition of an extra fact to nature is thus different from the dualism famously associated with René Descartes, which foundered on how "spirit" (or "thought") could be logically connected to the material world and exert influence on it. What Chalmers is postulating is an addition to the basic list of scientific principles that, when fleshed out, could take its place beside other basic laws, such as Newtonian (or Einsteinian) gravitation, Maxwell's equations of electrodynamics, and the quantum laws of atomic and molecular structure. These basic laws are not reducible one to another, nor need the principle of consciousness be reducible to any combination of the others.

Chalmers's idea can be distinguished from theories of consciousness as an *emergent* property of certain biological systems, a better known concept. Only when certain DNA sequences and corresponding proteins are aggregated together in cells (omitting, here, other essential molecules) does one see the phenomenon of life unfold. Various organs in a body have their specific properties because of their molecular organizations. One can describe such properties as "higher level" phenomena, dependent on the organization and structure of molecules, that "emerge" under specific conditions and have their own behavior. A heart beats. An intestinal system digests food. Lungs breathe. The overall system seems to be in charge. Without classifying himself explicitly as an "emergentist," the Nobel neuroscientist Roger Sperry saw the overall pattern of neuronal activity in the brain as taking over the direction of the constituent parts. If this overall pattern were constitutive of consciousness, Sperry argued, then our conscious thoughts could be seen as directing at least parts of the whole. Sperry used the simile of a wheel rolling down a hill: was it best understood by attempting to analyze the forces acting on each individual atom, or by seeing the forces to which the whole structure of the wheel responded? To attempt the former, let us suggest, might be simply "reinventing the wheel."

Emergent properties exist all the way down the scale of structures to the quantum states of individual molecules. The molecular orbitals of electrons in various chemical compounds confer new properties—especially in proteins such as enzymes, but even in very simple molecules. Consider the water molecule. The philosopher John Searle is fond of referring to the impression we have of the "liquidity" of water as something novel—even as a metaphor for how mental phenomena related to consciousness are not predictable from physical facts. For me this metaphor is not meaningful, since it already presupposes a subjective observer to detect "liquidity." A more significant scientific fact about water is the partially asymmetric structure of the molecule, which aids in the formation of beautiful snowflakes, but especially in the formation of miniature molecule-scale aggregates in water itself, under certain conditions. As we saw in chapter 3 (cf. figure 15), the dispersal of such microcrystalline water molecule aggregates, conforming to the Second Law of Thermodynamics, is the driving force in the formation of "hydrophobic bonds" within or between protein molecules. This is one thing that makes water (which is most of our body mass) such an ideal medium for living cells.

Another example. The chemical structures of the bases in DNA involve molecular orbitals of electrons that encompass the six- and five-atom rings. When these ring structures are participants in a double helix there is "stacking" of these wide orbitals, a real physical interaction of adjacent base pairs, that stabilizes the helix in addition to the hydrogen-bonding between the complementary bases in each base pair. This base-pair stacking effect is a property of DNA which reflects its complexity as a double-strand molecule and, in that sense, is an emergent property. (See figure 6 and figure 16.)

In my own view the idea of consciousness as an emergent property of the central nervous system is compelling. It is one that is consistent with the concept of location in such a system as a prerequisite for a unique individual consciousness to exist. It is also an idea that fits current scientific knowledge of developmental biology and the stages from embryogenesis to the adult organism, with a nervous system centered in the brain. We have already discussed the hypothesis of consciousness as a continuum through evolution of animals with more and more complex central nervous systems, having brains as distinctive centers. Here I would like to add another dimension to the continuum. It is personal consciousness—awareness—as an *outcome of development:* from fertilization of the egg through the stages of embryogenesis to the fully formed fetus and, finally, to the birth of the infant, even including, for humans, the steps toward full selfhood offered by parental care, family life, and acquisition of language. Real consciousness—self-awareness—certainly does not start at the moment of conception or in the first spherical assembly of embryonic stem cells. We know from experiments in other mammals that one such stem cell can be inserted into a different nascent embryo and end up having descendant cells in unpredictable parts of the future fetus and postnatal animal. (If the objective is to genetically engineer the single cell before insertion into the second pro-embryo, to create a new genetically altered line of animals, it is a matter of luck if the inserted cell contributes to formation of the reproductive organs of the postnatal animal.)

Without reviewing all the complex stages from the blastula and the inner cell mass to the more developed embryo, it suffices to say here that there are many of them and that the homeotic (HOX) genes have important roles to play, including in the development of the neural tube, which is the source of eventual brain parts and, in connection with appearance of the rudimentary vertebral column, of the spinal cord. There are complex steps in the outgrowth of the various subdivisions of the brain. By the end of gestation this anatomy is a precondition for the beginnings of sensory experience, of memory, and of the basic pleasures humans experience in being cared for (or, we hope, rarely, the pain of not having adequate care).

Against this background the controversies over embryonic stem cell research seem strangely irrelevant. If the stem cells concerned originate from in vitro fertilization, or, à la "Dolly the sheep," from replacement of the nucleus of an egg cell by the nucleus of an adult "somatic" (body) cell—say, from the skin—these cells, now maintained in cell culture, are not automatically destined to become cells of a central nervous system. Even if a single such cell, inserted into the initial sphere of embryonic cells, contributed to a nervous system, it would be simply a matter of statistics, not of predestina tion—as with attempts to target the future reproductive system. To ascribe any individual consciousness to such cells is meaningless. They are "at the extreme end" of the additional dimension of the consciousness continuum, which means zero consciousness. One end of the first dimension, the evolutionary dimension, would likewise be a zero, representing the first single-cell life form on this planet capable of self-replication. Some theorists of the origin of life, such as Richard Dawkins, and others, would extend this dimension even further back, to systems of replicating molecules. I believe that no one would ascribe consciousness to such systems.

Embryonic stem cell cloning with a somatic cell nucleus could be done with a medical patient's own skin cells, and a resultant cell clone might be treated with the appropriate growth factors to produce new insulin-producing cells for a patient with advanced diabetes, or heart-muscle cells for someone with a serious coronary condition—possibly even cells for transplantation into the brain of someone with an advanced phase of Parkinson's disease. The great advantage would be that because *all* the genetic information in the cloned cells came *from the patient* there would be no problem of immune rejection of the transplanted cells. More work needs to be done on the growth factors and cell-signaling molecules needed to convert the stem cells into the requisite tissue types, but this is a frontier in basic research that many laboratories, worldwide, are eager to explore.

I would like now to return to the main topic of this chapter and cite the ideas of someone who is arguably the greatest living authority on this subject.

Antonio Damasio, the noted clinical neurologist and humanist neuroscientist, has put forward a theory of consciousness in a recent book, *The Feeling of What Happens*. There he argues that the image of an object, as directly experienced or imagined, can produce a feeling of real or potential comfort or discomfort, pleasure or pain, tranquility or danger. In a fundamental sense we are aware of ourselves as organisms with a desire to survive and avoid injury. This is a root of consciousness and a sense of self—the identification of our own bodies. Tactile sensations and feelings within our bodies are more basic than the more "sophisticated" senses of vision, hearing, and even olfaction and taste. In an earlier book, *Descartes' Error,* Damasio proposed that we are constantly receiving, in our waking state, "reports," as it were, from various parts of our bodies. If put into words, these might come across as "things are OK here" or "this doesn't feel so good" or "something really needs to be done about this" or, again, "this is pretty nice" or "this is *wonderful, beautiful;* if only it could continue." From this basic set of feelings we generalize to reacting to all kinds of freshly presented images of objects, or remembered images, adding our intelligent capacity to draw connections concerning the significance for ourselves as living organisms.

I would propose, in conclusion, that human consciousness serves as the basis for *empathy,* starting from the unavoidable belief that others must have some feelings like ours, and, by our communicating this *shared* experience, as a source of love and much in our deeper spiritual lives.

SEVEN

MOLECULAR BIOLOGY AND THE LIMITS OF MOLECULAR EXPLANATION

I recall bringing my older daughter and a friend to a meeting connected with a summer program for placing American teenagers in families overseas; on the way we stayed at an inn in New England.* Across the road from the inn was a wide open field. When others had gone to bed I walked into this field and looked up at a dazzling sky full of stars that can be seen only dimly in New York City. Each of those points of radiance located a solar furnace, some perhaps bathing planets like Earth in light from which their biospheres extracted energy. All that I could see up there were countless sources of electromagnetic radiation.

I shifted my reverie then to the organs in my own body with which I was taking in this scene. The lenses and retinas of my eyes, optic nerves, and visual regions of the cerebral cortex were presenting the images of the starry night to my consciousness: all intricately fashioned products of evolution, all lineal descendants of those constellations of precursor molecules in primitive seas and tidal pools of the prebiotic earth. Light was not only the bearer of energy; it was the palette from which the incessantly changing picture of the world was painted, the world that I *saw*. The word *light* also connotes for us understanding, lucidity. Vision, of all the senses, has most enabled

* The program was the Experiment in International Living.

us to decipher and understand the world. The structures of human knowledge, of evolution regarding itself, are as real, fragile, yet capable of propagation as the organs of our bodies. Mental structures, preserved in language, are the ultimate transformation of radiant energy.

Mental experiences are a late product of evolution. They depend on brains with sufficient excess capacity to interpose some conscious reflection between a perceived stimulus and a consequent muscular reaction. The frog snapping at a fly has, most of us would agree, little mental space for reflection. Yet repetition, which characterizes the life of instincts and reflexes, retains its fascination for our human imaginations. There is something magical in repetition, in the rhyme of poetry or the return of a theme in music. We are constantly seeking pattern in the world, in our lives, and the basis of most pattern is repetition. Science itself starts from an awareness of regularities in nature. The greatest satisfaction in science is to see the same unifying principle expressed in a great diversity of situations. Replication of form and the recognition of identity and nonidentity are molecular requirements for biological reproduction, are features of sexual activity in all relevant species, and are basic to the highest achievements of human emotion and thought.

The power of mathematics grows from the recognition of identities. Close cousins of identity are ratio and precise proportion. The latter are what resound in musical harmonies and in the vibratory nature of sound itself. It is a curious fact that the sense of hearing—the other sense that tells us most about the world and, like sight, is able to give us intense aesthetic pleasure—responds to vibrations of all types of objects, carried through wave motions of the atmosphere. The oscillations and vibrations of things! The dance of natural phenomena is accompanied by myriad versions of these motions, of energized particles moving in potential energy fields. Particles in the quantum world of atoms and mountains shaking from the shifting of tectonic masses follow a similar mathematical music. The propagation of energy from sources, such as in all forms of electromagnetic radiation, is wave-like. Quantum mechanics and wave mechanics are the same discipline.

The human ability to grasp these ideas is inherent in the structure of the brain, genetically endowed like all structures of our bodies. But the content of our brains in thoughts and fantasies results from the ceaseless impinging upon us of energy in the waves of sensory signals that we perceive. Through language we organize these experiences and interact with the experiences of other people. The interaction permits the building of structures that form parts of our specifically human culture—the texts of books in libraries, the machines and other artifacts of our civilization—structures that go far beyond what is blueprinted in our genetic endowments and require vast amounts of energy beyond the maintenance of living things on the planet.

All this creation of new forms requires the expenditure of energy. Energy from the sun—directly, or stored in fossil fuels or the water cycle. Nuclear power is our only nonsolar energy source. Drawing on these energy sources, we are creating molecule-scale disorder as we create order in the form of new structures. The heat released in combustion, or in metabolism, is the most obvious version of such molecular disorder passed on to our environment. Far from flouting the Second Law of Thermodynamics, life's engine exploits the law at every turn. The shedding of disorder goes on in parallel with the elaboration of new structures, biologic and cultural. Those structures that persist can undergo further refinement: a process complementary to the drift toward simpler, more random patterns of matter and energy throughout the universe.

In the early part of the last century Niels Bohr enunciated a principle he called "Complementarity," to express a paradox in quantum mechanics. The paradox had one formulation in Heisenberg's "Uncertainty Principle": the more precision achieved in the measurement of the velocity of a particle, the more uncertainty in its location. The end of precise, billiard-ball mechanics. The end of prediction of future worlds to any desired degree of precision. Each preferred his own formulation. Measurements of particle speed and particle position were complementary—as were the timing of an event and the amount of energy exchanged in the event. There were complementary ways of viewing events in the quantum world that could not be simultaneously realized with exactitude. An analogy suggests itself in the living world, for which the physicists may forgive me. The emergence of structures with the potential to survive, and the subsidence into disorder of those that do not, offer us two views of Nature, but the subject, for both biologist and physicist, is a single one: the mystery of energy made substantial as matter.

It doesn't seem a "given" that raw matter, a fraction of the chemical elements, had the capacity to bring into being a "you" and a "me." *Subjective* states of mind are natural, with individual brains digesting sensory experience. But why such richness, such transcendent feelings of beauty, of love, of religious awe? In the human sphere, the texture of our own experiences and actions, we see the *limits of explanation* in molecular terms. We exist. We live. We are the actors in this drama, be it tragic or comic. And our actions can leave an imprint on the lives of others, and even on future generations. Science, in this instance molecular biology, offers detailed, fine-grained explanations, but these are not necessarily available at the moment that we must *act*.

Genes scarcely ever survive as isolated units; they are clustered in the species genome, and survival is at the level of the individual of the species.* Genes mutate singly, as a rule, and their persistence can be measured in the gene pool of a species population, but an adaptive trait also increases the survival probability of the associated genetic traits that collectively define the species . . . the genetic framework (species genome) within which the individual genes mutate and individual organisms live and die. (I use the term *species genome* in the same sense that a species phenotype can be recognized

---

* Viruses are a special example, usually a small set of genes, that cannot reproduce without the help of living cells, in any case.

as a basis for classification.) Altruistic behavior is not merely clannish, benefiting the genotypes of siblings and first cousins; it operates in the interests of the species genome.

The Second Law of Thermodynamics and the Principle of Natural Selection are, for the purposes of the biologist, logically complementary. In its statistical form the Second Law states that any (closed) system will spend most time in those macro-states for which the ensemble of energy-equivalent micro-states is the largest in number; that is, the system will move, in time, toward the most probable states, which happen (because of the dense configurations of micro-state ensembles for such an outcome) to be what we would see as disorderly or random ones.[1] No one who has seen a teenager's room need doubt the validity of the Second Law. Natural Selection states that when evolution moves toward more complex—orderly—forms, it does so because of selection of those genotypes that are the fittest to survive. Predominance of the "most probable" states and the "fittest" life forms might appear to be tautological propositions, but both are real statements about the natural world. Thermodynamics would be sterile without the existence of real physical systems. Life with its genetic mechanisms manifestly exists; it has sprung from matter itself, and its space-time distributions, as inexorably as the birth and death of stars.

Biological form is undecipherable unless placed in the framework of evolution. George Gaylord Simpson, an outstanding American evolutionary biologist of the last century, once advocated the viewpoint that biology is a more encompassing science than physics or chemistry. His argument: biology presupposes an understanding of the principles and phenomena of the sciences as they pertain to atoms and molecules; it adds to them the principles and phenomena of larger organized systems that are evolution-selected. From this union of the sciences we can see how a principle of physics, the Second Law of Thermodynamics, and a principle of biology, Natural Selection (i.e., Survival of the Fittest), are logically complementary—as I've already stated. Evolution defies disorder—

or, better, it places disorder at the service of life, exploiting the trend toward microdisorder for the sake of macrostability. Even if we were to concede, for the sake of argument, that Survival of the Fittest has its tautological aspects, then so does the principle that molecular aggregates tend to spend most time in the most probable (numerous) states. The two "tautologies" are balanced across eons in evolution. From this tension has arisen life as we know it, the actions of statesmen and mobs, the understanding of scientists and the images of poets.

By the end of the last century the shared efforts of scientists worldwide had culminated in a detailed description of how information is transferred from DNA into protein and of the ingenious ways in which gene expression is regulated. Always the emphasis was on the *minimum* number of molecules that codetermine an outcome, and the proof of the pudding has been the reconstitution of the system in question, from purified molecules, in vitro ("in glass"—test tubes). There is something astringent, austere, and rather beautiful about this quest for the grail of simplicity in biology. The touchstone is "Does it work?"—a question that can be answered by a properly designed in vitro experiment and plenty of hard work in identifying and purifying the components of the system. The premise is that a cell consists of a finite number of such microsystems, which, if combined together, will be a perfectly valid cell once more. No élan vital à la Henri Bergson, no embryo-directing "entelechy" as postulated, over a century ago, by Hans Driesch, no vital force need be supplied. Since the last-mentioned sorts of things have not been observed in in vitro systems, and have no place in the cell's total information store in its DNA, they can be relegated to the dustbin of philosophical romanticism.

Molecular structure and cellular architecture are what count here, and the overall organization of the cell at a particular moment. The instructions in DNA seem to be most essential to what makes a particular cell what it is. The iconoclastic Richard Dawkins has lobbied strenuously for making the story of evolution the story of DNA itself, advancing step by step up the ladder of natural selection. There is just one problem. DNA by itself is naked and helpless. It needs a cell to decode it and be informed by it. New DNA is always freshly minted inside of cells (except when a molecular biologist makes it in vitro). "Every cell arises from another cell," said Virchow, the great nineteenth-century pathologist. The reason for this (true) statement is clear: a cell passes on to its mitotic offspring enough of the basic molecular machinery for reading DNA and following its instructions so that things can go on as usual. The fertilized ovum, of course, has its own maternally derived machinery for this purpose. Thus, paralleling Dawkins's time line of gene-bearing molecules stretching back to the origins of life, there is an unbroken sequence of other molecular contents of cells back to the earliest precellular thrashings about of macromolecules attempting to do it all by themselves. What rivets our attention here is the emergence, over time, of a cellular apparatus that enables the cell to express only a limited set of its total genes at any given time. This limited expression may be in response to environmental signals or to the imperatives of embryonic development and differentiation in a multicellular organism.

The regulation of gene expression! This has been a grand theme of biological research in the past century, but only in the last few decades have experimental methods been available to make substantive inroads on the problem. The molecular cloning of precise segments of DNA, including DNA reverse-transcribed from RNA, has been the key, opening the way to the sequencing of thousands of DNA bases in many genes and to the identification of elusive proteins and their production at the wish of the scientist. The reconstitution of transcriptional control phenomena in the test tube, or in test cells grown under laboratory conditions, is now routine. Through these approaches, far more intricate and powerful than this brief sketch would suggest, a picture of surpassing evolutionary ingenuity has begun to emerge.

dependent on the renewal of their grasses, just as we are on the modern farms of Iowa. The bee is dependent on the flower, and vice versa. Bees could not exist without the complex language of bee communication, the capacity for which is part of their hereditary makeup. The beehive itself can be regarded as a kind of organism. This is Lewis Thomas on the subject of bees:

> Bees live lives of organisms, tissues, cells, organelles, all at the same time. The single bee, out of the hive retrieving sugar (instructed by the dancer: "southsoutheast for seven hundred meters, clover, mind you make corrections for the sundrift") is still as much a part of the hive as if attached by a filament. Building the hive, the workers have the look of embryonic cells organizing a developing tissue; from a distance they are like the viruses inside a cell, running off row after row of symmetrical polygons as though laying down crystals. When the time for swarming comes, and the old queen prepares to leave with her part of the population, it is as though the hive were involved in mitosis. There is an agitated moving of bees back and forth, like granules in cell sap. They distribute themselves in almost precisely equal parts, half to the departing queen, half to the new one. Thus, like an egg, the great, hairy, black and golden creature splits in two, each with an equal share of the family genome.
>
> *(The Lives of a Cell)*

We humans communicate too, in more elaborate ways, and our language may fairly be said to be the basis of civilization. The use of language was first made possible by changes in the genetic makeup of our ancestors, and in all probability the benefits of language caused further, reinforcing hereditary changes in our brains. Nevertheless, raise a human being, even in imagination, in isolation from other humans, and this capacity for language will have no outcome. The contact with other people is vital for language use itself to develop and thus for contact with the reservoir of human knowledge. Let us focus for a moment on where this reservoir is. First of all it is in our brains—wired in, so to speak, by what we have learned since birth—for us so little of what we know is already encoded in us at the moment of our conception, as compared with the bees. But we have also transferred a vast amount of what we know to written texts, pictures, diagrams, and, not least of all, to tools, edifices, and machines. If we forgot how to build a house, and there was a house still standing somewhere, we could study it, and learn again to build it, as the Renaissance architects studied the remains of Greek and Roman buildings. Now I want you to think for a moment of all these human creations, including the specific patterns in our brains, as physical structures that were absent from the earth during most of its history and have sprung up upon its surface since the advent of the human species. These are *truly evolved structures,* just as much as the skeletal structures of mammals or the structure of a beehive. They have sprung from the crust of the earth, through the endless succession of organic molecules that first formed in primordial seas, through the sequence of the generations of living things, and, finally, owing to the combined efforts of the human beings who have lived before us. In fact, if you think hard about it, it is impossible to regard human culture except as part of evolution, as the latest stage in the process that began with the association of the first molecules in the primordial soup.

Let us think about the brain for a moment. The anatomical structure is encoded by our genes. Does that mean that some thoughts are already in our genes? Some analysts of the psyche, such as Jung, have given a partial affirmative to this. Based on what we know about genes, it is more likely that the capacity for certain ways of thinking, perhaps especially for linking certain types of thoughts with certain emotions, is present from birth. This part of human nature is itself the end product of the tinkering and refining of the evolutionary process—the selection of the kind of brain that would work in the world (mostly of jungle and cave, the world of our hunter and gatherer ancestors, when the anatomy of the brain was laid down in something like its modern form—which may be one of our problems). But probably a more important part of the actual physical structure of our brains at the micro-level, where the synaptic connections are wired, is the result of our experience, growing up in a human culture. We don't necessarily choose to learn all of what we learn; it is constantly being imprinted on us from birth, though the interaction with our developing feeling of having our own center, our own ego and sense of choice, is there almost from birth and grows stronger as we grow older. And, because we as individuals are at the center of our lives, we are identical with this process of sifting all the inputs, making new neuronal junctions, establishing a net of minute electrical connections that is uniquely our own as we go through life. Here is how the great neurophysiologist Charles Sherrington described the electrical activity of the brain as it passes from sleeping to waking. For the sake of description he is imagining that each brain cell is a tiny point of light, rather like a great city seen from the air at night:

> Should we continue to watch the scheme we should observe after a time an impressive change which suddenly accrues. In the (cortex) which has been mostly darkness spring up myriads of twinkling stationary lights and myriads of trains of moving lights of many directions. It is as though activity from one of these local places which continued restless in the darkened main-mass suddenly spread far and wide and invaded all. The great topmost sheet of the mass, that where hardly a light had twinkled or moved, becomes now a sparkling field of rhythmic flashing points with trains of traveling sparks hurrying hither and thither. The brain is waking and with it the mind is returning. It is as if the Milky Way entered upon some cosmic dance. Swiftly the head-mass becomes an enchanted loom where millions of flashing shuttles weave a dissolving pattern, always a meaningful pattern though never an abiding one; a shifting

harmony of subpatterns. Now as the waking body rouses, sub-patterns of this great harmony of activity stretch down into the unlit tracks of the stalk-piece of the scheme. Strings of flashing and traveling sparks engage the lengths of it. This means that the body is up and rises to meet its waking day.

(*Man on His Nature*)

Is this process determined by the cause-and-effect chemistry of the individual nerve cells, which, as biochemists like to think, might be analyzed as precisely as a transistor or, with due allowance, perhaps a computer chip? Or does the overall pattern of mental activity control the individual cell? To say the latter is not to fall into unbridled mysticism—for atoms and molecules can always only follow the rules according to the structures in which they are placed. Remember "Sperry's wheel," in which the atoms are like any other atoms, but, as the wheel rolls downhill, their movement relative to each other is determined by the structure of the wheel, of which the atoms have no knowledge.

Here we come to something crucial. Granted that structures like wheels and brains control how the atoms within them move, that our adult brain can be said to control, in the conscious flow of thought, the individual nerve cells and the atoms within them, how does the initial structure get there? We have the answer already. The structures, both brains and wheels, are there because of the evolutionary process, sustained by solar energy over billions of years. This is also what makes a brain different from a computer, which is a human construct, rather like a wheel. The brain stands as the end product of the myriad selection events that have left their record in the chemistry of the genes. The history of the whole species, and also of earlier stages in evolution, is imprinted in our genes, from which arises the infant anatomy of our brains. And it is the accumulated history of human culture, interacting with this anatomy, that is the basis of our adult minds. In a broad perspective each structure in evolution is more complex and more embracing, and each is the point of departure for creation of still more complex structures with new potentialities.

Thomas Huxley's grandson, Julian Huxley, a well-known biologist of the century just passed, once asked where, in the explanations and exact predictions of science, could one find a new "touchstone for ethics." Where in all of what we have discussed might we find Huxley's touchstone for ethics? After all, you might argue, even if the organisms that we know are the outcome of a complex evolutionary process, even if the activity of our brains is much more elaborate than operation of a single nerve cell, itself composed of a vast number of molecules, doesn't it all add up in the end to a physical process ruled by deterministic laws? Where then is the conscious choice of our actions? There are two quick answers to this—one is very old, one is relatively new. The old answer was given perhaps best by Socrates, as he was about to drink the hemlock. (He is making fun of someone who would say [This is Socrates talking]:)

The reason why I'm now sitting here is that my body consists of bones and sinews, and the bones are hard and separated from each other by joints, whereas the sinews, which can be tightened and relaxed, surround the bones, together with the flesh and the skin that holds them together; so that when the bones are turned in their sockets, the sinews by stretching and tensing enable me somehow to bend my limbs at this moment, and that's the reason why I'm sitting here bent in this way; or again, by mentioning other reasons of the same kind for my talking with you, imputing it to vocal sounds, air currents, auditory sensations, and countless other things, yet neglecting to mention *the true reasons:* that Athenians judged it better to condemn me, and therefore I in my turn *have judged it better* to sit here, and *thought it more just* to stay behind and submit to such penalty as they may ordain. Because, by the dog, these sinews and bones would long since have been off in Megara or Boeotia, impelled by their judgement of what was best, had I not thought it more just and honorable not to escape and run away, but to submit to whatever penalty the city might impose.

(Plato, *Phaedo,* trans. David Gallop; emphasis added)

In fact, ever since the Greeks first framed the atomic theory of matter, it was clear that atoms moving in the void did not remove the necessity for human beings to make choices, and, if necessary, agonize over those choices. For how can I know where the atoms are when the moment of choice is at hand? The molecular explanation for our actions is not available for the decision-making process. One need not even adopt a relativistic or Humean view of causality as a human category imposed on nature; the causal explanations of science at the molecular level are simply not relevant to action as a conscious psychological event.

There is a newer answer to this age-old question as well. It is that when we speak, or attempt to speak, of a physical process as extremely complex as mental activity, cause and effect in the billiard ball sense is scarcely applicable. One needs a new science, such as psychology, to handle what is going on, and everyone knows that psychology is not an exact science. I would like to suggest that this is not because of any deficiencies on the part of psychologists, or their science, but because the subject matter itself defies precise description—by its very complexity it contains an element of the unpredictable, even a dramatic clash of creative and destructive desires, as Freud saw it. Nowadays we can even conceive of building a computer for which the designer could not predict what responses it would give, if the machine were supplied with a specific set of questions. In fact, I remember that the little chess computer I had on my desk some years ago didn't always play the same game twice.*

---

* My guess now is that its program included something equivalent to a random number generator or, alternatively, some instruction to "learn" from the previous game(s).

But is unpredictability a touchstone for ethics? I would like to argue that something more fundamental than unpredictability is at work in evolution. To understand what this is we need to consider once again the small miracle of the bee and the flower. Neither can survive without the other. The evolution of these creatures is certainly not obvious from a scrutiny of the building blocks of matter. We might say that these creatures are "emergent structures," freshly created by evolution and then sustained. Sustained because they work, they fit together. It needn't have turned out that way. The creation of these creatures—of all organisms, in fact—represents a sort of "breakthrough," an actualization of a potential in matter, but certainly not the sole or dominant trend in cosmic events. Remember the Second Law of Thermodynamics and the tendency to disorder in things. Disorder seeps into the lives of living creatures all the time. It is perhaps particularly poignant when the survival of the species is at stake. The California condor is such a species, and it seems to me that the following example of disorderly behavior in the face of species extinction has a vaguely human ring:

> The only California condor egg known to have been laid this breeding season was knocked off a cliff by the parents, who were fighting.
>
> A team of specialists trying to save the huge, endangered birds watched from half a mile away as the four-inch egg smashed on the rocks below the condors' cave and the embryo was eaten by ravens. The condors, the specialists said, were apparently battling over which of them should take care of the egg.
>
> "There was real musical chairs going on in the nest," said John Ogden, co-director of the program to save the vultures, which have wingspreads up to 12 feet.
>
> "One of them would sit down on the egg and the other would come in and try to push the first one off," he said Thursday. "They would jab each other in the face and really get physical."

. . . At one point, Mr. Ogden said, the quarreling became so heated that both birds took to the air, squawking and flailing at each other with wings and talons, leaving the egg unattended.

"This went on for hours," Mr. Ogden said. "They were so absorbed, no one was incubating."

(UPI, Ventura, California, March 5, 1982)

The appearance of some species in evolution is what might be called a "one-time" thing, depending on a specific outcome at a specific time and place. Stephen Jay Gould argued strongly for this. Not only are all creatures mortal, all species have to gamble occasionally for their existence, and the outcome is not a foregone conclusion. The geneticist Theodosius Dobzhansky defined a species as an interacting pool of genes: through sexual mating the different genetic possibilities within the framework of the species are constantly being tested, and those combinations of genes that work well together, enabling individuals to survive and have offspring, will contribute most copies to the gene pool. Richard Dawkins, by way of contrast, tends to see genes as discrete units, knocking others out of the competition to persist into the next generation. But there are good reasons to regard the particular assemblage of genes in a species gene pool as adapted to one another, rather the way that bees are adapted to flowers. The beautiful fit of different parts within an organism has always been the delight of biologists and of amateur naturalists, nature photographers, and artists. There is now some evidence from modern studies of evolution that each species may represent a sort of preferred way of doing things, a kind of temporary resting point in the process of evolution where Nature has, as it were, said: this seems to be a pretty good design, maybe worth keeping around for a while. The technical term for this idea is "punctuated equilibrium"—proposed by Niles Eldredge and Stephen Jay Gould—in which there are rapid transitions between species, with fewer intermediary forms to leave fossil traces. One of the constraints on Nature trying out all conceivable varieties of genes together may be the limitations on a large animal growing from an embryo into an adult. To pass through all the intermediate stages, forming the right tissues and organs, may depend on there being a limited number of good fits between the products of genes that are active during development. This factor in the evolution of species has been called by Stephen Jay Gould and others "developmental constraint."

For the molecular biologist the notion of precise fits within the contents of a single living cell is always at hand. Enzymes have to fit together with other enzymes; the whole cell is a piece of molecular architecture in space and time that can be upset only too easily by random mutations—valuable as these can be in the laboratory for dissecting the vital processes of the cell. *Molecular cooperation* is a good term for what is going on. Cooperation is evident throughout biology, between different cells, between organisms, and between the organism and its environment. The last should not be overlooked. The full realization of the genetic potential in an organism requires interaction with the environment. I would like to offer a somewhat complicated example of this. Think for a moment of the play and fantasy life of a young child: it springs not from the inherited structure of the child's brain, but from the contact of the child's mind with the world around it. These experiences are then organized by the child into a web of fantasy and inner storytelling that is unique to each child and his or her particular experience of the world. This is how the vital feeling of "being at home in the world" comes into being. Every organism interacts intimately with its environment. The plant fits into the environment of the soil; at the most basic level, it has long been known to biochemists that the structure of water itself is uniquely fitted for the maintenance of life, and of the biosphere.

We are part and parcel of the world that we live in. *This is our touchstone for ethics.* We are part of a process that has been going on on the Earth for a long time, and it is now our responsibility to take care of this process and the new dimension that has been realized by the creation of human culture.

I would argue that there is no room for a mind-body dualism, in the sense that Descartes defined it, because we simply are and have our existence in organisms that are the end product of the vast duration of evolution. The consciousness of "being me" is simply the experience of *being* the organism rather than "looking in" from the outside. Our being rooted in the world, in this very concrete and specific sense, runs counter to the Platonic notion of our souls descending into the world at the start and ascending out at the finish of our lives. A modern Catholic theologian, Karl Rahner, has argued that, in fact, the material aspect of the human organism is an essential prerequisite for there to come into existence a truly human life of the spirit.

Our own mortality as individual organisms is not in question. What must concern us as caretakers of the evolutionary process is the mortality of human life as a whole, of civilization, and of the entire intricate life-support system on this planet as we know it. The ethical choices are thrust upon us by the fragility of this system as it now coexists with human culture. I will not detail here the long-range environmental problems created by our industrial societies. To cite an extreme threat, the massive destruction, human suffering, and potential, persisting atmospheric changes caused by all-out nuclear war, it is the ultimate irony of hydrogen-based nuclear weapons that, like the Greek myth of Phaeton, they disastrously bring into the sphere of the earth the energy source of the sun, which itself has nourished life on this planet through the chlorophyll of green plants and thus into the food chain of all living organisms. But even without the possible species-suicide of nuclear war, there is armed conflict, abject poverty, and bad human health in many parts of the world, as well as risks to the survival of other species. And we, even if in more fortunate socioeconomic circumstances, cannot predict the ill-fated events that might come our way. In sum, there is tragedy enough in human existence. Our response to tragedy must be to reach out to offer help to our fellow human beings, to practice Reverence for Life, and to reinforce our effort to find—and to help *create*—meaning, love, and beauty in life. The heroism of life comes from its being fragile. All the best things that we strive for can be destroyed—are mortal, just as we ourselves are. I am convinced that evolution on this planet, in its human phase, has been partly a heroic effort to construct and preserve a kind of life that is worth living—but it is not an effort guaranteed to succeed, and only with hope, intelligence, and hard work can we sustain it.

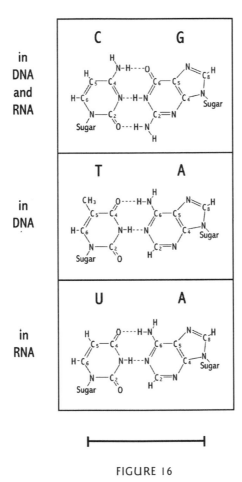

FIGURE 16

Chemical structures of DNA and RNA base-pairs. The dotted lines represent hydrogen bonds. Three bases, designated C, G, and A, are common to both DNA and RNA. In RNA "U" takes the place of the "T" of DNA in the "base quartet." It is easy to see here how an RNA might be copied from a DNA by base-pairing (cf. figure 9). The scale bar represents 10^{-6} millimeter.

PROLOGUE: MOLECULAR BIOLOGY FROM THE
VIEWPOINT OF A PRACTITIONER

1. DNA, as a single strand molecule, consists of a "backbone" of alternating sugar and phosphate (phosphorus/oxygen) groups, with a "base" attached to each sugar. There are four bases, symbolized as A, T, G, and C, that are single- or double-ring structures (see figure 16). The rings themselves are composed of carbon and nitrogen atoms. There are some functionally important oxygens bonded to three of the ring structures, as well as significant nitrogens—also in three cases (not the same). There are also some hydrogen atoms to occupy otherwise empty bonding sites; some can participate in base-pair interactions (figure 16). The "basic" property is due to nitrogen atoms.

1. THE GENETIC POINT OF VIEW

1. Beadle's and Tatum's trailblazing experiments were done in the mold *Neurospora,* a kind of rudimentary eukaryote, but they worked with the stage in the life cycle after spore germination, when there are only single copies of genes present, facilitating mutational analysis. These kinds of experiments were subsequently repeated in bacteria—which have single gene copies.

2. THE LOGIC OF THE CELL

1. Dedicated as he was to the European traditions of pure organic chemistry and biochemistry, Chargaff was at first skeptical of some of the developments in molecular biology, with its genetic and macromolecular components. He was not an immediate convert to the double-helical model of DNA structure, nor to the experimental logic of the genetic code, although base-pairing was the linchpin for both discoveries.

2. My enzyme assay (test) was simple. Just put a radioactive version of the activated methionine methyl donor in an appropriate solution in a test tube with methyl-deficient tRNA, add my protein preparation from my frozen cells, at whatever stage of purification, incubate at a healthy temperature (such as human body temperature, which the bacteria, found in the human gut, were used to), and finally precipitate (make insoluble) the whole mixture of proteins and RNA with a suitable acid. Then, after washing the precipitate on a small filter to remove the traces of unused methyl donor, dry the filter, and measure the retained radioactivity in the appro-

priate counting instrument. Controls were to perform an incubation with normal RNA in place of methyl-deficient RNA or to use the latter RNA but not perform the incubation. Seeing the printout from the counting instrument was satisfying for someone with a quantitative bent.

3. I could apply these new findings in the work that I carried out at Rockefeller. A friend of mine, Dr. Jack Goldstein, introduced me to the laboratory of Lyman Craig, where tRNA fractionation into different amino acid–specific subspecies was carried out by a method called "countercurrent distribution." The possibility existed that the methylated and non-methylated forms of phenylalanine-specific tRNAs might separate under these conditions; one could then recover them and test them individually. This was the case, and an analogous situation existed for other amino acid–specific tRNAs. (There were reliable sources from which I could obtain synthetic RNAs such as poly U and others.) In the course of time I also learned to appreciate Lyman Craig's skill in volleying at the net during the weekly tennis doubles match on the Rockefeller court, in which I was invited to join.

4. The helper virus for Rous virus could be any member of a class of chicken leukemia viruses, widespread in chicken flocks in this country and other parts of the world, going back at least to the beginning of the century—unable to infect mammalian cells—more on this and related subjects in chapter 4. The Schmidt-Ruppin strain of Rous virus, which I had used in the course at Cold Spring Harbor, was something of a special case. It did not require helper virus in chicken cells—a further genetic development—and could thus grow in chicken cells, and yet also infect rodent cells, but without producing any progeny virus in the latter case, although the cells were rendered tumorigenic. In the rodent the virus was capable of forming tumors only when inoculated into a newborn, thus circumventing the animal's potential immune capability to reject virus-infected cells.

3. FROM CELL TO ORGANISM

1. It should be mentioned that in the fly there is maternally derived information residing in molecules of the egg cytoplasm and cell membrane that has a crucial role in embryonic patterning of the larva and the future adult fly.

2. A recent estimate is that there are approximately thirty-two thousand genetic loci. (A genetic locus represents a region of DNA that has the potential to code for protein without many interruptions by "stop"

codons, which would normally terminate translation, or by transposable genetic elements, which would also interrupt normal transcription as well as translation.) Many loci could have a potential for encoding more than one protein, as a result of two processes not so far discussed. These are 1. multiple *splicing* patterns for arriving at an mRNA from the original RNA produced by transcription of a locus and 2. potential feasibility of beginning translation of an mRNA as a base triplet sequence in three ways that, in effect, split the triplets into alternative translational reading frames. (1) reflects a normal process in molecular cell biology of eukaryotes: RNA transcribed from a locus can lose some base sequences, known as "introns," by a complex catalytic process known as "splicing." The mRNA product can have more than one form, if alternative introns are removed. (2) is simply a consequence of the fact that there are three different ways in which triplets can be "read off" from a plain sequence of bases in an RNA molecule (before stop codons are taken into consideration). The effect of (1) and (2) is thus to open up the possibility that a single genetic locus may, in fact, encode more than one protein. This increases the potential amount of effective genetic information in the human genome well beyond the raw number of genetic loci that can be detected. J. Darnell, personal communication.

3. It is a familiar idea that a chemical reaction is driven in a certain direction by the release of heat, raising the temperature of the surroundings. If the surroundings are fluid (liquid or gaseous), a rise in temperature is defined as an increase in the average kinetic energy of the molecules there—i.e., more vigorous motion. This results, at the molecular level, in more disorder.

4. The "methylation" of tRNA, which I found to increase efficiency in the function of some tRNAs in protein synthesis, could also be explained by the probable formation of a hydrophobic bond of the tRNA to the ribosome, where it could fit better into a precise pocket in performing its function of delivering the next amino acid to the nascent protein chain.

5. For the sake of completeness, one should also mention the role of so-called van der Waals interactions between molecules due to spontaneous/induced electrical polarity (mediated by orbiting electrons of the component atoms). These can be a factor in intermolecular binding, if sizable areas of the molecular surfaces are involved in the interactions.

4. ABNORMAL CELL GROWTH IN CANCER

1. The immunological test I decided to use was known as "complement fixation," in which a blood component, called "complement," binds to

antigen-antibody complexes and one measures what is left over by a standard method. This gives a fairly quantitative picture of how much antigen is present in a sample.

2. The probe was made by reverse transcription of Rous virus RNA, allowing the DNA product to bind to mutant Rous RNA lacking the transforming gene, and saving that part of the DNA that did not enter into the hybrid helices for potential helix formation with cell-derived DNA strands.

3. At this time the term "oncogene" is not used to designate a gene that is necessarily directly linked to an endogenous retrovirus. Rather, oncogenes are regarded as genes in normal cells that can have significant roles in causing cancers, if mutated or inappropriately expressed. The latter can, of course, be the case if such a gene is acquired by a retrovirus through genetic recombination (as in Rous sarcoma virus having acquired the *src* gene). The achievement of Huebner's and Todaro's proposal was to focus attention on specific genes, residing in chromosomes, with the dual potential, under abnormal conditions, to drive cells toward a cancerous transformation and to become inserted into retrovirus genomes. What was not so clear at that time (1969) was the very significant functions that the naturally evolved versions of these genes have in normal cell physiology.

5. HUMAN GENOMES AND HUMAN BEINGS

1. The main part of this chapter is based on a book review written by the author, for the July-August 1999 issue of the *Hastings Center Report*, of a book published by the Tuskegee University Publications Office entitled *Plain Talk About the Human Genome Project: A Tuskegee University Conference on Its Promise and Perils . . . and Matters of Race,* edited by Edward Smith and Walter Sapp (Tuskegee, Ala.: 1997). Where not otherwise attributed, quotations and citations in this chapter are from this book.

7. MOLECULAR BIOLOGY AND THE LIMITS OF MOLECULAR EXPLANATION

1. The term *micro-state* is used here to denote one of the ways that energy can be absorbed into a molecular system, that leaves the molecular *structure* intact. There are essentially three classes of ways that this can happen: (1) the movements of the molecules through space, (2) the vibrations between the different atoms that form a given molecule, and (3) the rotations of the atoms, singly or as clusters, leaving the overall molecular bonding struc-

bases

A base in DNA or RNA is the chemical structure that, in 4 versions, in various combinations with these other bases, carries the distinctive genetic information of that DNA or RNA molecule. See note 1 of the prologue, with accompanying figure (figure 16), and figure 6 in chapter 1.

biosphere

The sum total of living organisms on this planet, with the various environmental conditions that support their lives.

cell

The fundamental life-form, containing DNA as genetic material, and the molecular apparatus needed to synthesize the *proteins* that are encoded in the DNA, through which the cell is able to extract energy and molecular "raw materials" from its environment—all contained within a cellular outer *membrane,* consisting of lipid (fatty) and protein constituents.

central nervous system

The brain and the *spinal cord.*

chromosomes

The molecular structures, consisting of DNA wrapped in *proteins,* that carry the genetic information in *cells* with *nuclei.*

cloning

The production, by biological methods, of genetically identical copies of DNA molecules, *viruses, cells,* or organisms—with the implication that an indefinite number of copies can be made.

codon

A unit of the *genetic code:* a sequence of 3 RNA *bases* in *m*RNA that codes for 1 *amino acid* in the *protein* encoded by that mRNA and is recognized by base-pairing to the 3 bases of the *anticodon* of the appropriate *t*RNA.

cytoplasm

For a *cell* of the *eukaryote* type, this is a term for all the cell contents that lie between the cell *nucleus* and the cell's outer *membrane.*

dendrite

A filamentous extension of a nerve cell (*neuron*) that is available for contact by an *axon* from another neuron, with potential reception of a signal from that neuron at the contact point (synapse).

DNA

A polymer molecule with a backbone of alternating sugar (deoxyribose) and *phosphate* groups, with 1 of the 4 fundamental DNA *bases* bonded to each sugar. DNA most often exists as a pair of molecules twisted around each other in an elongated double-helix structure, with hydrogen bonds between pairs of bases facing each other across the axis of the two helices. See figure 16 with note 1 of the prologue and also figure 6. DNA is the central bearer of genetic information in cells and organisms.

endocrine

(As in endocrine gland, endocrine hormone, endocrine system). The term designating aspects of the hormonal systems that involve secretion into the bloodstream of specific *hormone* molecules that carry signals to those cells in other parts of an organism that have appropriate molecular receptors on their surfaces (see *hormone*).

enzyme

A molecule, almost invariably a *protein,* which can act as a catalyst in accelerating the rate of a biochemical reaction. Enzymes, like other proteins, are encoded in *genes,* and there are thousands of them in living *cells,* vital to the processes that go on. These include expression of genetic information, synthesis of new molecules, generation of energy from nutrients, and specialized functions of tissues and organs. Words denoting enzymes usually end in the three letters *ase.*

equation of state

The term in physics for the mathematical equation, expressing basic laws of physics, that, when specific values of certain variables are known, determine exact values of other variables, thus defining the "state" of the system as of a given time.

eukaryote

An organism with genetic information that is carried in *chromosomes,* which are in the *nuclei* of *cells.*

*f*MRI

Magnetic resonance imaging of tissues or organs that shows the functioning of the component cells through changes in the image over time. For example, in the case of the brain, such changes can reflect increased or decreased blood flow in a particular area. This, in turn, can be interpreted as greater or diminished neuronal activity in the area.

focus

In *tissue culture* of animal *cells,* in flat cell *monolayers,* a concentrated group of cells that grow more densely, with an altered cell shape, which can reflect infection of a single cell by a tumor *virus,* causing abnormal extent of cell growth and alteration in cell shape. A few infectious tumor virus particles can produce an equivalent number of foci when spread out over a cell-culture plate with a sufficient number of cells to produce areas of flat cell monolayer as well as discrete foci where an infection event occurred.

genes

The functional units in which genetic information is carried. A gene gives rise to a distinct gene product (in general, a *protein* molecule) that has a specific role in the life of the *cell* or organism. This process can be termed gene "expression."

genetic code

The molecular units, consisting of triplets of DNA (or RNA) *bases,* in which the 20 *amino acid* constituents of *protein* molecules are represented (encoded). Since, with 4 bases, there are 64 possible triplets, there is more than one triplet for each amino acid. A very few triplets have no amino acid counterpart and represent "stop" signals for the *translation* process.

genome

The complete set of genetic information in a *cell* or organism—in its most detailed form, as the exact sequence of DNA *bases* in the DNA.

genotype

The characteristic genetic features of an organism that distinguish it from other species and from other members of the same species. The most precise version of the genotype is the *genome.*

growth factor

A molecule, most often a *protein,* that binds to a specific receptor molecule on the exterior of a cell *membrane* and stimulates *cell* growth and cell division, resulting in an increase in cell number. Particular growth factors are specific for certain cell types that display appropriate receptors.

hormone

A generic term for a molecule released from a specialized *cell,* which targets another cell type and evokes a specific (molecular) response. Best known are *endocrine* hormones, which are produced by specialized endocrine glands and travel through the bloodstream until they encounter matching receptors on the surfaces of cells elsewhere in the body.

HOX genes

These are *genes* that have a fundamental role in the formation of distinct body parts in animals, from insects to mammals (cf. chapter 3). The original term is *homeotic* genes, meaning genes that, when mutated, can cause abnormal expression of a particular body part, as a duplication, in the wrong position. Homeotic genes have a characteristic DNA sequence as part of their information; this is often diagrammed in a "box," known as the "homeobox"—hence the term HOX.

in vitro

In molecular biology, the term used to denote a laboratory procedure in which some constituents of a *cell* population are removed from the cells and tested in a suitable container (the classic one being a test tube—the root meaning of *in vitro* is "in glass"). A test could be designed to detect an *enzyme* activity, or a cooperative process involving several enzymes and other molecules.

lawn

(of bacteria)Contiguous bacterial colonies (of the same strain of bacteria) that cover the surface of the nutrient agar in a petri dish and serve as an ideal support system for creation of *bacteriophage plaques.*

membrane

The outer membrane of a *cell* is what separates the cell and its contents from the molecular environment exterior to the cell. For animal cells such membranes have a double layer (bilayer) of fatty acid molecular constituents, which restricts transit of water-soluble molecules and ions. *Protein* components are inserted into the bilayer or arranged on the exterior or interior surfaces at various points, and have distinct roles to play in cellular functioning. There are also systems of internal membranes within eukaryotic cells, e.g., the membrane of the cell *nucleus.*

monolayer

In *tissue culture* of animal *cells:* cells that have grown on a flat surface, of plastic or glass, covered with a nutrient liquid medium, so as to cover the entire surface in a geometry that is "one cell deep" (between the surface of the culture vessel and the medium above it). Monolayers are ideal for visualization of *plaques* or *foci* caused by infection with very dilute preparations of animal viruses.

*m*RNA

Stands for "messenger RNA," which is the product of *transcription* of DNA, and carries the genetic information from particular *genes* to the *ribosomes,* where *translation* into equivalent *proteins* can take place.

myc

An *oncogene,* originally found in the RNA of the avian *retrovirus* MC29 (which induces a form of leukemia in chickens). *myc* is widely represented in many species, has normal roles to play in organisms, and, in some genetic versions, has been implicated in some human cancers.

natural selection
In Darwin's theory of evolution, the process by which certain chance variations, in physical form, functioning, or behavior, of some individuals of a species, are instrumental in the survival of more next-generation descendants, and still further generations, from those "varieties" of the species, thus establishing the original variations as lasting, hereditary features of the species. In Darwin's theory this was the way that related, new species eventually arose in Nature. Also known as "survival of the fittest."

neuron
A general term for a nerve cell, forming part of the *central nervous system* (brain, *spinal cord*) or the peripheral nervous system (rest of the body).

neurotransmitter
A molecule that carries a signal from the *axon* of a *neuron* to a receptor molecule on another cell (another neuron or a different cell type receptive to such a signal). The neurotransmitter crosses a short extracellular space to reach its receptor, such that, in the microscope, the neuronal axon would appear to be in direct contact with the other cell.

NIH
The abbreviation for the National Institutes of Health, located near Washington, D.C. The NIH is a large set of government-financed laboratories and is the major site of basic medical science research directly supervised by the federal government.

nucleus
In the anatomy of a *cell* of the *eukaryote* class, the *membrane*-enclosed, interior subspace that contains the *chromosomes* and attendant structures of that cell.

oncogene
A cellular *gene* that, if mutated or improperly expressed, can cause a *cell* to move on to the path of being a cancer cell. Those *retroviruses* that can cause cancers in experimental animals have, in general, either acquired a copy of a cellular oncogene (hence: *viral* oncogene) or can integrate, as viral *DNA,* into cell *chromosomes* so as to activate, on a statistical basis, the improper expression of a cellular oncogene. Oncogenes have important roles in normal cell functioning.

optic cortex (primary)
A region on the surface (hence *cortex*) of the cerebrum in the brain that records the immediate transmission of the train of neuronal signals originating in the retina(s) of the eye(s). Hubel and Wiesel were Nobel laure-

ates for showing, in experiments in primates, that particular classes of visual stimuli (e.g., moving versus static) caused distinctive patterns of neuronal activity in the primary optic cortex.

organelle

A specialized structure within a *cell,* such as the cell *nucleus,* that may have its own boundary *membranes.*

Parkinson's disease

A disorder of the brain involving poor coordination of bodily movements as well as involuntary movements. The physiological basis of the disorder involves a deficiency in the amount of the *neurotransmitter* dopamine in crucial parts of the brain. Drug treatment addresses this deficiency.

PET *scanning*

PET is an abbreviation for *p*ositron *e*mission *t*omography. This stands for the imaging, as though in flat sections, of a part of the body—achieved by injecting into the blood a harmless chemical that is labeled with an isotope that emits positrons. Positrons are the positive-charge analogs of electrons and are annihilated with the first electrons they (soon) meet to form high-energy X-rays, which are detected by a scanner—showing blood circulation there.

phenotype

The biological form and functioning of an organism, including outward behavior, that is genetically based. Can be compared with the *genotype,* which is the organism's actual genetic constitution. The phenotype is what in fact leads to more successful or less successful survival in *natural selection.*

phosphate

A chemical grouping of a phosphorous atom ringed by 4 oxygen atoms, 1 or 2 of which can routinely be used, in biological systems, to form "ester bonds" to other molecules or parts of molecules—2 such bonds hold each sugar molecule in the sugar-phosphate "backbone" of DNA and RNA. See also *phosphorylation.*

phosphorylation

In biological systems, the process, characteristically catalyzed by *enzymes,* in which a *phosphate* group is linked—by an ester bond—to another molecule, in some cases a *protein,* in terms of the discussion in this book.

plaque

In bacterial systems: a small circular "hole" or clear patch in an even "*lawn*" of bacteria on nutrient agar, representing an original infection of one *cell* by a single *bacteriophage* particle. Starting with a dense suspension of viral particles, one can make serial dilutions (tenfold or more at each step) until a small volume of one of the last dilutions makes individual plaques that are countable—thus permitting a calculation of the original virus concentration. (A plaque results from viral progeny of a single infected cell spreading to neighboring cells in a circularly symmetric fashion that is repeated over time.) An analogous process takes place when cell-killing animal viruses are diluted before infection of a cell *monolayer.*

prokaryote
A single-cell organism without a *nucleus,* with DNA that is not in a *chromosome* structure. Bacteria are typical prokaryotes.

protein
A large molecule consisting of a chain ("polymer") of *amino acids,* connected to each other by a characteristic type of chemical bond made possible by shared structural features of the amino acids. Because there are 20 amino acids that participate in forming proteins, the latter are a very diverse and versatile class of molecules, with many biological functions, especially as *enzymes* in cells. Proteins also function as components of cell *membranes* and of *chromosomes,* as cellular *growth factors* (and their specific receptors), as *oncogene* products, in essential ways in *neurons,* and in a multitude of other ways that make living organisms what they are (one more example: as the major constituent of muscle tissue).

quantum theory
In physics and chemistry: the theory that energy exists, and is exchanged, in discrete amounts—as opposed to there being a "sliding scale" or continuum of energy in a system. By contrast, in the mathematical formulation of the theory, material particles, especially subatomic particles, do not have precise locations and velocities at the same moment in time—instead having a "wave-like" character—while wave-radiation (electromagnetic radiation, e.g., light or heat) can be seen as having a precise, particulate character, consistent, in its being a manifestation of pure energy, with the first statement above.

quanta
The specific, and precise, quantities of energy that are characteristic of particular molecular, atomic, and subatomic systems, which can be in part released by such systems in the form of discrete amounts of electromagnetic radiation. Conversely, corresponding amounts of energy can be absorbed by such systems, in discrete quantities, that restore the systems to their original states.

quantization of energy

According to the *equations of state* of *quantum theory,* the amounts of energy that are characteristic of a physical/chemical system can be expressed as discrete, mathematically predictable values. These values often apply to definite energy levels that a system can attain by acquisition or release of energy as *quanta* in the course of chemical reactions or directly via effects of electromagnetic radiation.

ras

An *oncogene* originally isolated from a *retrovirus* (of mouse derivation) found in association with a *rat* sarcoma in a laboratory context (virus injected experimentally into rat). Cellular *ras* genes exist in many species and have important roles in *signal transduction pathways.*

retrovirus

A type of *virus* that carries its genetic information in the form of RNA, yet, upon infection of a new host *cell,* converts this information to a DNA molecule, through the agency of the viral *enzyme, reverse transcriptase,* that is carried in the virus particle. The viral DNA, with the help of another viral enzyme, is then integrated into the host cell DNA in a *chromosome.* From this site new viral RNA is produced and processed, by *transcription* and *translation* with host cell RNA *polymerase* and *ribosomes,* to yield fresh RNA and *proteins* for new viral particles, which bud from the cell's outer *membrane,* carrying a piece of this membrane studded with selected viral proteins as the external surface of each new virus particle. Retroviruses do not kill their host cells; they can bud from such cells indefinitely, unless an immune response from a host organism is vigorous enough to intervene (*then* causing death of the host cell). (Note: the agent of human immunodeficiency disease, HIV, is a retrovirus.)

reverse transcriptase

The *enzyme* carried in a *retrovirus* that converts the viral genetic information, which is in an RNA form, to an equivalent DNA in the infected *cell* — thus reversing the usual flow of information in *transcription,* in which the *base* sequence in the DNA of one of a cell's *genes* is copied into an equivalent sequence of RNA bases.

ribosome

A submicroscopic particle in a *cell* on which the RNA copy of genetic information in the DNA of a *gene* undergoes *translation* to yield a *protein,* which is the expressed form of the gene. Ribosomes consist of RNA and protein molecules bound together in a distinct geometry: there is a "small subunit" and a "large subunit" that together form the complete particle.

RNA

A polymer molecule that, like DNA, has a backbone of alternating sugar and *phosphate* groups, in this case with the sugar being ribose, and that also has 1 of 4 fundamental RNA *bases* bonded to each sugar (see figure 16, of note 1 of the prologue). Of 4 bases, 3 are identical in DNA and RNA. RNA is generally single stranded, though parts of a molecule may loop back to form a double-helical segment of limited size.

RNA polymerase

An *enzyme* that copies the genetic information in a DNA *gene* into the form of a single-strand RNA molecule.

Rous sarcoma virus

The infectious agent isolated by Peyton Rous at the Rockefeller Institute, New York City, from a solid tumor of a chicken. The work was published in 1911. In general, derivatives of the original virus of Rous are not infectious for other species and require a "helper virus" even to reinfect chickens (see chapter 2 and select bibliography and further reading for this chapter).

Second Law of Thermodynamics

An important principle of physics that governs the use and exchange of energy (the First Law is that of Conservation of Energy). The Second Law states that in a "closed system," where energy is held constant, there will be a trend, over time, toward overall states of the system that are associated with a larger number of "molecularly detailed" micro-states. This trend could be accomplished, for example, by there being a spread of available energy, in the form of heat, more uniformly throughout the system, coinciding with a statistically larger variety of molecular motions for the system as a whole. This can be regarded as an increase in "disorder" at the molecular level.

Signal transduction pathways

In cell biology: the sequences of molecular interactions by which signals pass from one set of molecules to another—often from one part of a *cell* to another. Specific chemical reactions and interactions are frequently involved—*enzyme*-mediated *phosphorylation* of *proteins* represents one class of such phenomena. There can also be precise adhering of one protein to another, as exemplified by the protein products of *ras* oncogenes. Signal transduction, in defined steps, can proceed from the cell *membrane* to the *nucleus* and then to specific *gene* locations on *chromosomes,* causing gene expression.

Spinal cord

In *vertebrate* animals: the main trunk of nerve fibers and nerve cells that proceeds from the brain down the spinal column and connects the brain, with the head region, to the rest of the body via nerve fibers (*axons* or *dendrites*).

src

The first *oncogene* to be clearly identified, *src* is the tumor-inducing gene of Rous *sarcoma* virus, which was "captured" by an avian precursor virus, of the *retrovirus* class, from a normal chicken *cell.*

telomerase

An *enzyme* that has the function of restoring DNA sequences at the ends of DNA in *chromosomes.*

telomeres

The repetitive DNA sequences at the ends of *chromosomes,* which tend to become shorter with each *cell* division, a process that is counteracted by *telomerase* — otherwise the loss of telomere sequences can have the result that cell division is inhibited.

tissue culture

The procedure in research in which *cells* from an animal are grown on a flat surface covered with a nutrient medium, under sterile conditions. When initially seeded at a low density, the cells can eventually form a *monolayer.* Cells that are cultured may be derived from fresh embryos, from specimens of tumors, or from "established cell lines" that have attained the ability to grow indefinitely in culture. (The second category may grow beyond monolayer densities.)

transformation

In *tissue culture:* transformation of *cells* frequently refers to altered cell shapes and increased cell population, encountered when cells are infected with a *virus* carrying a mutated or aggressively expressed *oncogene,* or are exposed to carcinogenic chemicals or radiation. Over half a century ago "transformation" was used to denote the experimental alteration in properties of bacteria caused by exposure of bacterial cells to purified DNA (see select bibliography and further reading for chapter 2: Avery, MacLoed, and McCarty). This meaning is still valid, also for exposure of cells of *eukaryote* origin to DNA.

transcription

The process, catalyzed by an *enzyme, RNA polymerase,* in which the DNA *base* sequence of a *gene* is copied into an equivalent RNA base sequence in a single-strand RNA molecule (*m*RNA) that can then direct the synthesis of the corresponding *protein,* which is the ultimate product of gene expression.

translation

The molecular steps by which an *m*RNA, bearing *genetic code* information from a *gene,* is the source of a particular *amino acid* sequence in a freshly synthesized *protein.* A new sequence language for a gene's DNA sequence: hence, *translation.* The steps involve participation of *ribosomes, t*RNA molecules (representing each amino acid), and a considerable reservoir of *enzymes* (some enzymatic activity seems to be intrinsic to the ribosome).

*t*RNA

An abbreviation for "transfer RNA," a term for a group of relatively small RNA molecules, each of which can be recognized by an *enzyme* dedicated to attaching to it a particular *amino acid* and also possesses a sequence of 3 RNA *bases* that can pair with a complementary triplet of bases in *m*RNA; tRNAs are thus pivotal in the process of *translation.*

ultracentrifuge

An instrument for spinning a solution of a (large) molecule in such a strong centrifugal field that the molecule will move toward the bottom of its container (tube) at a rate that can be monitored by an appropriate optical device. This permits calculation of the molecular weight of the molecule.

vertebrate

An animal with a central backbone and an overall skeletal structure displaying considerable symmetry from left side to right side ("bilateral symmetry").

virus

A submicroscopic structure consisting of genetic information in the form of DNA or RNA, enclosed in *protein*(s) that are virus encoded (and in some cases, such as *retroviruses,* an outer envelope derived from the infected cell *membrane,* with viral protein(s) embedded in it). Viruses cannot reproduce by themselves; they require some of the molecular apparatus of living *cells,* which they infect, to supply the "building blocks" and "tools" (*amino acids,* DNA or RNA *bases* and accompanying sugar-phosphate structures, molecules and molecular structures essential for *translation,* etc.).

X-ray diffraction
The research procedure in which a submicroscopic structure, generally a molecule or a very small set of molecules, is prepared in a geometrically organized fashion (especially as a crystal of pure molecules, or a preparation of fibers—as in the case of DNA—oriented in a particular way) and then subjected to irradiation with X-rays of defined wavelength. This allows repetitive aspects of the sample's structure to deflect ("diffract") the X-rays at particular angles that are registered by an electronic detector. It is then possible to calculate back to probable molecular structures, even to the level of individual atoms, in the sample under study.

SELECT BIBLIOGRAPHY AND FURTHER READING

I. THE GENETIC POINT OF VIEW

Beadle, G. W. "Biochemical Genetics: Some Recollections." In *Phage and the Origins of Molecular Biology*. Ed. J. Cairns, G. S. Stent, and J. D. Watson. Dedicated to Max Delbrück on the occasion of his sixtieth birthday. Cold Spring Harbor, N.Y.: Cold Spring Harbor Laboratory Press, 1966.

Benzer, S. "Adventures in the rII Region." In *Phage and the Origins of Molecular Biology*. Ed. J. Cairns, G. S. Stent, and J. D. Watson. Dedicated to Max Delbrück on the occasion of his sixtieth birthday. Cold Spring Harbor, N.Y.: Cold Spring Harbor Laboratory Press, 1966.

Dobzhansky, T. *Genetics and the Origin of Species*. New York: Columbia University Press, 1937.

Mayr, E. *The Growth of Biological Thought: Diversity, Evolution, and Inheritance*. Cambridge: Harvard University Press, 1982.

2. THE LOGIC OF THE CELL

Avery, O. T., C. M. MacLoed, and M. McCarty. "Studies on the Chemical Nature of the Substance Inducing Transformation of Pneumococcal Types: Induction of Transformation by a Desoxyribonucleic Fraction Isolated from Pneumococcus Type III." *Journal of Experimental Medicine* 79 (1944): 137–158.

Brenner, S., F. Jacob, and M. Meselson. "An Unstable Intermediate Carrying Information from Genes to Ribosomes for Protein Synthesis." *Nature* 190 (1961): 576–581.

Crick, F. H. C. "The Genetic Code: III." *Scientific American* 215.4 (1966): 55–62.

Hanafusa, H., T. Hanafusa, and H. Rubin. "The Defectiveness of Rous Sarcoma Virus." *Proceedings of the National Academy of Science: USA* 49 (1963): 572–580.

Hershey, A. D. and M. Chase. "Independent Functions of Viral Protein and Nucleic Acid in Growth of Bacteriophage." *Journal of General Physiology* 36 (1952): 39–56.

Lake, J. A. "The Ribosome." *Scientific American* 245.2 (1981): 84–97.

Losick, R. "Summary: Three Decades After Sigma." In *Mechanisms of Transcription.* Cold Spring Harbor Symposia on Quantitative Biology, volume 63. Cold Spring Harbor, N.Y.: Cold Spring Harbor Laboratory Press, 1998.

Monod, J., J.-P. Changeux, and F. Jacob. "Allosteric Proteins and Cellular Control Systems." *Journal of Molecular Biology* 6 (1963): 306–329.

Rich, A. and S. H. Kim. "The Three-Dimensional Structure of Transfer RNA." *Scientific American* 238.1 (1978): 52–62.

Rous, P. "A Sarcoma of the Fowl Transmissible by an Agent Separable from the Tumor Cells." *Journal of Experimental Medicine* 13 (1911): 397–411.

Temin, H. and H. Rubin. "Characteristics of an Assay for Rous Sarcoma Virus and Rous Sarcoma Cells in Tissue Culture." *Virology* 6 (1958): 69–688.

Watson, J. D. *The Double Helix: A Personal Account of the Discovery of the Structure of DNA.* New York: Atheneum, 1968.

3. FROM CELL TO ORGANISM

Birnbaumer, L. "Receptor-to-Effector Signaling Through G Proteins: Roles for ßγ Dimers as Well as ∝ Subunits." *Cell* 71 (1992): 1069–1072.

Capecchi, M. R. "*Hox* Genes and Mammalian Development." In *Pattern Formation During Development.* Cold Spring Harbor Symposia on Quantitative Biology, vol. 62. Cold Spring Harbor, N.Y.: Cold Spring Harbor Laboratory Press, 1997.

Ericson, J., J. Briscoe, P. Rashbass, V. van Heyningen, and T. M. Jessell. "Graded Sonic Hedgehog Signaling and the Specification of Cell Fate in the Ventral Neural Tube." In *Pattern Formation During Development.* Cold Spring Harbor Symposia on Quantitative Biology, vol. 62. Cold Spring Harbor, N.Y.: Cold Spring Harbor Laboratory Press, 1997.

Fantl, W. J., D. E. Johnson, and L. T. Williams. "Signaling by Receptor Tyrosine Kinases." *Annual Review of Biochemistry* 62 (1993): 453–481.

Gilbert, S. F. *Developmental Biology,* pp. 180–185, 242–248. Sunderland, Mass.: Sinauer, 1997.

McEachern, M. J., Krauskopf, A., and E. H. Blackburn. "Telomeres and Their Control." *Annual Review of Genetics* 34 (2000): 331–358.

Meier, P., A. Finch, and G. Evan. "Apoptosis in Development." *Nature* 407 (2000): 796–801.

Quirk, J., M. van den Heuvel, D. Henrique, V. Marigo, T. A. Jones, C. Tabin, and P. W. Ingham. "The *Smoothened* Gene and Hedgehog Signal Transduction in *Drosophila* and Vertebrate Development. In *Pattern Formation During Development*. Cold Spring Harbor Symposia on Quantitative Biology, vol. 62. Cold Spring Harbor, N.Y.: Cold Spring Harbor Laboratory Press, 1997.

Roeder, R. G. "Role of General and Gene-Specific Cofactors in the Regulation of Eukaryotic Transcription." In *Mechanisms of Transcription*. Cold Spring Harbor Symposia on Quantitative Biology, vol. 63. Cold Spring Harbor, N.Y.: Cold Spring Harbor Laboratory Press, 1998.

Hunter, T. "Protein Kinases and Phosphatases: The Yin and Yang of Protein Phosphorylation and Signaling." *Cell* 80 (1995) :225–236.

Hunter, T., and B. M. Sefton. "Transforming Gene Product of Rous Sarcoma Virus Phosphorylates Tyrosine." *Proceedings of the National Academy of Science: USA* 77 (1980): 1311–1315.

Schindler, C. and J. E. Darnell. "Transcriptional Responses to Polypeptide Ligands: The JAK-STAT Pathway." *Annual Review of Biochemistry* 64 (1995): 621–651.

Wodarz, A. and R. Nusse. "Mechanisms of Wnt Signaling in Development." *Annual Review of Cell and Developmental Biology* 14 (1998): 59–88.

4. ABNORMAL CELL GROWTH IN CANCER

Baltimore, D. " RNA-Dependent DNA Polymerase in Virions of RNA Tumor Viruses." *Nature* 226 (1970): 1209–1211.

Brugge, J. S. and R. L. Erickson. "Identification of a Transformation-Specific Antigen Induced by an Avian Sarcoma Virus." *Nature* 269 (1977): 346–348.

Collett, M. S. and R. L. Erickson. "Protein Kinase Activity Associated with the Avian Sarcoma Virus SRC Gene Product." *Proceedings of the National Academy of Science: USA* 75 (1978): 2021–2024.

Der, C. J., T. G. Krontiris, and G. M. Cooper. "Transforming Genes of Human Bladder and Lung Carcinoma Cell Lines Are Homologous to the RAS Gene of Harvey and Kirsten Sarcoma Viruses." *Proceedings of the National Academy of Science: USA* 79 (1982): 3637–3640.

Hayward, W. S., B. J. Neel, and S. M. Astrin. "Activation of a Cellular *Onc* Gene by Promoter Insertion in Avian Leukemia Virus-Induced Lymphoid Leukosis." *Nature* 290 (1981): 475–480.

Huebner, R. J. and G. J. Todaro. "Oncogenes of RNA Tumor Viruses as Determinants of Cancer. *Proceedings of the National Academy of Science: USA* 64 (1969): 1087–1094.

Hunter, T. and B. M. Sefton. "Transforming Gene Product of Rous Sarcoma Virus Phosphorylates Tyrosine." *Proceedings of the National Academy of Science: USA* 77 (1980): 1311–1315.

Moodie, S. A. and A. Wolfman. "The Three Rs of Life: Ras, Raf, and Growth Regulation." *Trends in Genetics 10* (1994): 44.

O'Donnell, P. V., E. Fleissner, H. Lonial, C. F. Koehne, and A. Reicin. "Early Clonality and High-Frequency Proviral Integration Into the *C-MYC* Locus in AKR Leukemias." *Journal of Virology* 55 (1985): 500–503.

Parada, L. F., C. J. Tabin, and R. A. Weinberg. "Human EJ Bladder Carcinoma Oncogene Is Homologue of Harvey Sarcoma Virus *Ras* Gene." *Nature* 297 (1982): 474–478.

Temin, H. M. and S. Mizutani. " RNA-Directed DNA Polymerase in Virions of Rous Sarcoma Virus." *Nature* 226 (1970): 1211–1213.

Varmus, H. "An Historical Overview of Oncogenes." In *Oncogenes and the Molecular Origins of Cancer.* Ed. R. A. Weinberg. Cold Spring Harbor, N.Y.: Cold Spring Harbor Laboratory Press, 1989.

5. HUMAN GENOMES AND HUMAN BEINGS

Cavalli-Sforza, L. L. *Genes, Peoples, and Languages.* New York: North Point, 2000.

Smith, E. and W. Sapp, eds. "Plain Talk About the Human Genome Project: A Tuskegee University Conference on Its Promise and Perils . . . and Matters of Race." Tuskegee, Ala.: Tuskegee University Publications Office, 1997.

Watson, J. D. *A Passion for DNA: Genes, Genomes, and Society.* Cold Spring Harbor, N.Y.: Cold Spring Harbor Laboratory Press, 2000.

6. ON CONSCIOUSNESS

Chalmers, D. J. *The Conscious Mind: In Search of a Fundamental Theory.* New York: Oxford University Press, 1996.

Damasio, A. R. *The Feeling of What Happens: Body and Emotion in the Making of Consciousness.* New York: Harcourt Brace, 1999.

—— *Looking for Spinoza: Joy, Sorrow, and the Feeling Brain.* New York: Harcourt Brace, 2003.

Deacon, T. W. *The Symbolic Species: The Coevolution of Language and the Brain.* New York: Norton, 1997.

de Duve, C. *Life Evolving: Molecules, Mind, and Meaning.* New York: Oxford University Press, 2003.

Dennett, D. C. *Freedom Evolves.* New York: Viking Penguin, 2003.

Humphrey, N. *A History of the Mind: Evolution and the Birth of Consciousness.* New York: Copernicus, 1992.

Schacter, D. L. *Searching for Memory: The Brain, the Mind, and the Past.* NewYork: Basic, 1996.

Searle, J. *Minds, Brains, and Science.* Cambridge: Harvard University Press, 1984.

Sperry, R..W. "Mind-Brain Interaction: Mentalism, Yes; Dualism, No." *Neuroscience* 5 (1980): 195–206.

7. MOLECULAR BIOLOGY AND THE LIMITS OF MOLECULAR EXPLANATION

Einstein, A. "The World as I See It." *Ideas and Opinions.* Trans. Sonja Bargmann. New York: Three Rivers, 1954.

Mayr, E. *This Is Biology: The Science of the Living World.* Cambridge: Harvard University Press, 1997.

Miller, S. L., and L. E. Orgel. *The Origins of Life on Earth.* Ed. W. D. McElroy and C. P. Swanson. West Nyack, N.Y.: Prentice-Hall, 1973.

Orgel, L. E. "Evolution of the Genetic Apparatus: A Review." *Evolution of Catalytic Function.* Cold Spring Harbor Symposia on Quantitative Biology, volume 52. Cold Spring Harbor, N.Y.: Cold Spring Harbor Laboratory Press, 1987.

Plato. *Phaedo.* Trans. David Gallop. New York and Oxford: Oxford University Press, 1993.

Rahner, K. *A Rahner Reader.* Ed. G. A. McCool. New York: Crossroad, 1981.

Schweitzer, A. *The World-view of Reverence for Life.* Trans. C. T. Campion. London: Black, 1923.

Sherrington, C. *Man on His Nature.* Gifford Lectures, Edinburgh, 1937–38. Cambridge: Cambridge University Press, 1941.

Virchow, R. *Cellular Pathology.* New York: Dover, 1971 [1858].

TEXTBOOKS

Guttman, Burton, Anthony Griffiths, David Suzuki, and Tara Cullis. *Genetics: A Beginner's Guide.* Oxford: Oneworld, 2002.

Malacinski, George M. *Essentials of Molecular Biology.* 4th ed. Sudbury, Mass.: Jones and Bartlett, 2003.

Turner, P. C., A. G. McLennon, A. D. Bates, and M. R. H. White. *Instant Notes: Molecular Biology.* 2d ed. Oxford: BIOS Scientific and New York: Springer, 2000.

HISTORIES

Judson, Horace Freeland. *The Eighth Day of Creation.* Cold Spring Harbor, N.Y.: Cold Spring Harbor Laboratory Press, 1996.

Morange, Michel. *A History of Molecular Biology.* Trans. Matthew Cobb. Cambridge: Harvard University Press, 1998.

Olby, Robert. *The Path to the Double Helix: The Discovery of DNA.* Seattle: University of Washington Press, 1974; rpt. New York: Dover, 1994.

Watson, James D., with Andrew Berry. *DNA: The Secret of Life.* New York: Knopf, 2003.

INDEX